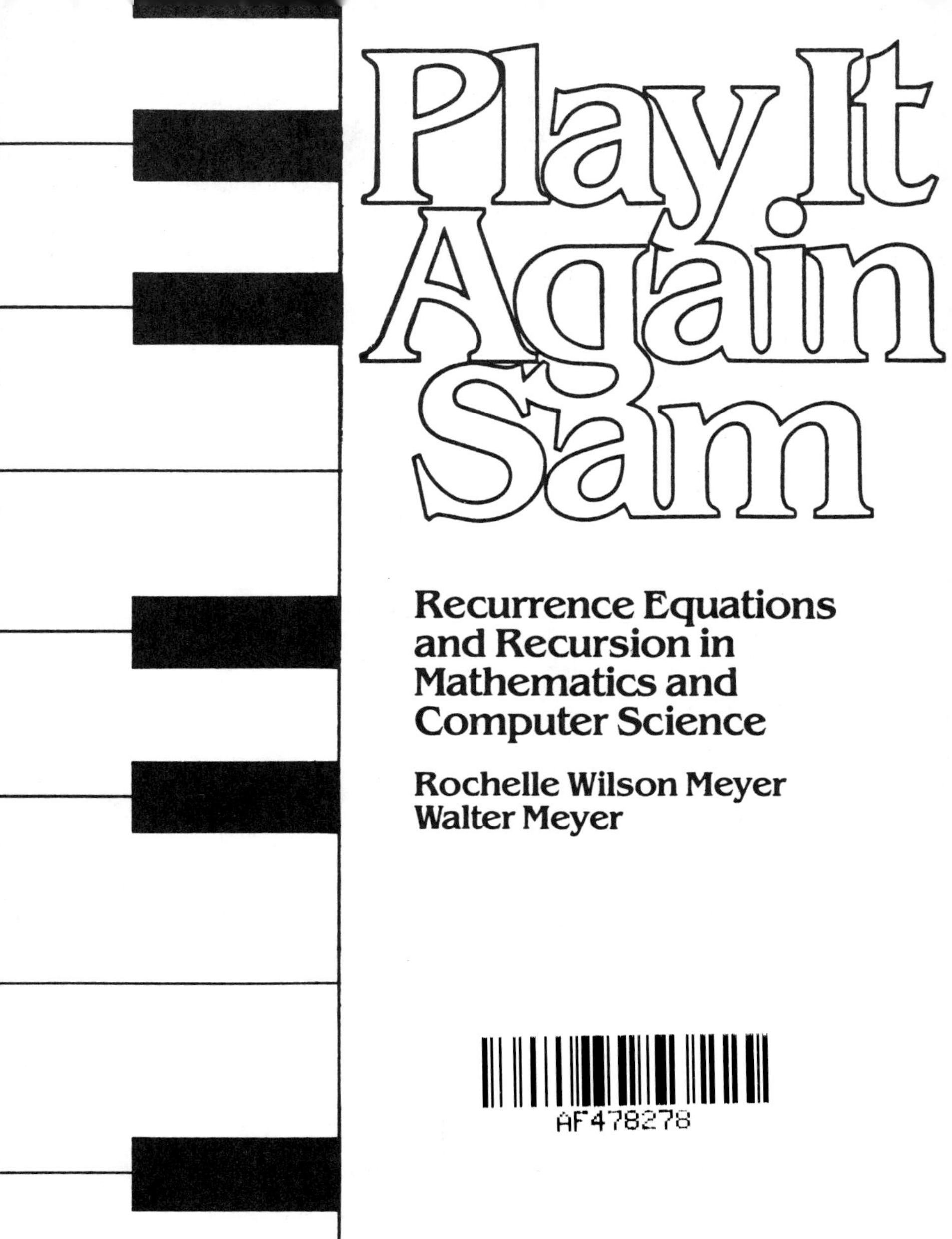

Play It Again Sam

Recurrence Equations and Recursion in Mathematics and Computer Science

Rochelle Wilson Meyer
Walter Meyer

COMAP's Explorations in Mathematics

HIGH SCHOOL MATHEMATICS AND ITS APPLICATIONS (HIMAP) PROJECT

The goal of HiMAP is to develop, through a community of users and developers, a system of instructional modules in high school mathematics and its applications that may be used to enhance teacher training at the secondary school level. The Project is guided by a national advisory board and an editorial board of mathematicians, scientists, and educators. HIMAP is funded by the National Science Foundation to the Consortium for Mathematics and Its Applications (COMAP), Inc., a non-profit corporation engaged in research and curriculum development in mathematics education.

COMAP Staff

Solomon A. Garfunkel	Executive Director
Laurie W. Aragon	Development Director
Philip A. McGaw	Production Manager
Roland Cheyney	Project Manager
Evelyn Moore	Copy Editor
Annemarie S. Morgan	Administrative Assistant
Dale Jolliffe	Production Assistant
Henry Beerman	Distribution Manager

HiMAP Project Staff

Solomon A. Garfunkel	Co-project Director
Joseph Malkevitch	Co-project Director

HiMAP Advisory Board

L.L. Clarkson	Texas Southern University
James Gates	National Council of Teachers of Mathematics
Irwin Kaufman	New York City Board of Education
Katherine P. Layton	Beverly Hills High School Beverly Hills, California
David Moursand	University of Oregon
Warren Page	New York City Technical College
Alex Rosenberg	University of California Santa Barbara
Gail Young	University of Wyoming

HiMAP Editorial Board

Beverly J. Anderson	University of District of Columbia
Robert B. Davis	University of Illinois
James T. Fey	University of Maryland
Gary Froelich	Bismarck Senior High School, Bismarck, North Dakota
John C. Howe	TRW Inc., Redondo Beach, California
James Kaput	Southeastern Massachusetts University
David Moore	Purdue University
John Riedl	Ohio State University

Contents

This material was prepared with the support of the National Science Foundation (NSF) Grant No. DPE8318104. Recommendations expressed are those of the author and do not necessarily reflect the views of the NSF or of the copyright holder. All correspondence should be addressed to:

Professor Joseph Malkevitch
COMAP, Inc.
60 Lowell Street
Arlington, MA 02174

Published by: COMAP, Inc., 60 Lowell Street, Arlington, MA
02174

ISBN 0-912843-17-9

Chapter One
Recurrence Equations

Section 1.1
What is a Recurrence Equation?

YOU HAVE PROBABLY SEEN QUESTIONS LIKE THIS ON TESTS, OR perhaps posed as brain teasers: Find the next number in the sequence

$$2, 4, 8, ...$$

Finding the answer depends on finding the pattern. In this case, the pattern is that each number is twice the one before. If we let $f(n)$ stand for the nth number in the sequence (counting the initial term as the 0th term), we could express this pattern as:

$$f(n) = 2f(n-1) \quad \text{for } n = 1, 2, 3, ...$$
$$f(0) = 2$$

Not all patterns are this simple. Can you express the pattern for

$$1/2, 2, 5, 11, 23, ... ?$$

The answer is

$$f(n) = 2f(n-1) + 1 \quad \text{for } n = 1, 2, 3, ...$$
$$f(0) = 1/2$$

The equations $f(n) = 2f(n-1)$ and $f(n) = 2f(n-1) + 1$ are examples of recurrence equations. The equations $f(0) = 2$ and $f(0) = \frac{1}{2}$ are called initial conditions because they describe how the sequence gets started.

Definition:

A recurrence equation is an equation which describes a sequence of numbers by showing how an arbitrary term of the sequence (except for the initial one) can be computed from one or more earlier terms. Often a recurrence equation is accompanied by an initial condition which states the first term, or the first few terms, in the sequence.

Here are some more examples of recurrence equations and initial conditions:

a. $f(n) = 3(f(n-1)^2) + 4$
$f(0) = 0$

b. $f(n) = f(n-1) + 2f(n-2)$
$f(0) = 1$
$f(1) = 2$

Question 1.1.1

Can you explain why it is appropriate to give two equations as the initial condition in example **b.**? Can you explain why it is unnecessary to give two equations as the initial condition for example **a.**? Can you give an example to show that giving two equations as the initial condition for case **a.** can be incorrect? Can you state a general principal for the appropriate number of equations in the initial condition?

The Flaw in Pattern-Guessing

A little known fact about pattern-guessing problems is that there are many possible answers. For example, the sequence we started with, 2,4,8, ..., can also be described by

$$f(n) = 2f(n-1) + (n-1)(n-2)$$
$$f(0) = 2$$

To verify this, let's substitute $n = 1$ and $n = 2$ into the formula:

$$f(1) = 2f(0) + (1-1)(1-2) = 2f(0) = 2(2) = 4$$
$$f(2) = 2f(1) + (2-1)(2-2) = 2f(1) = 2(4) = 8$$

What we have done is taken the expression for the old pattern, $2f(n - 1)$, and added an expression, $(n - 1)(n - 2)$ that evaluates to 0 for $n = 1$ and 2. Thus, we have a pattern (formula plus initial condition) that still fits the first three terms, 2, 4 and 8 and so it is just as good as our earlier solution. However, it does give a different prediction as to the next term, namely

$$f(3) = 2f(2) + (3 - 1)(3 - 2)$$
$$= 18$$

whereas, using the original recurrence equation gives

$$f(3) = 2f(2) = 2(9) = 16$$

Question 1.1.2

Do you see how to carry out this kind of trick and find a recurrence equation different from $f(n) = 2f(n - 1) + 1$ for the sequence ½, 2, 5, 11, ... ? If you can, then perhaps you can do it for any sequence at all. (In view of this, does it make sense to have such questions on tests?) A final challenge: suppose you want to find an alternate pattern for the sequence 2, 4, 8, ... which predicted 2.418 (or any other specified number d) for the fourth term. Can you tinker with the recurrence equation to do it?

An Application of Recurrence Equations

It is not very common in science or mathematics to start with a sequence of numbers and then look for a recurrence equation that describes it. It is more common to start with a problem that results in a recurrence equation and initial condition and then to draw some conclusions about the sequence that is described by it, as in Fibonacci's famous rabbit breeding problem.

In this problem, described by Fibonacci of Pisa in 1202, we assume rabbits live forever and that every month each breeding pair begets another pair consisting of a male and female who will become a breeding pair after two more months. If we have a newborn pair in month 0, how many will we have in succeeding months?

Let $f(n)$ denote the number of rabbit pairs after n months. We start with $f(0) = 1$. After 1 month there is still no reproduction so $f(1) = 1$. But after 2 months the rabbits we started with reproduce and we have $f(2) = 2$. At 3 months, all the $f(2)$ rabbits who were there at 2 months are still there (because they live forever), but in addition there is one newborn pair for each pair that is at least two months old. The number of pairs that is at least two months old is the number of pairs present 2 months ago, i.e. $f(1)$. Therefore

$$f(3) = \text{pairs still there} + \text{newborn pairs}$$

$$f(3) = \qquad f(2) \qquad + \qquad f(1)$$

The same kind of relationship is true for all further n. In other words,

$$f(n) = f(n-1) + f(n-2)$$

This recurrence equation, together with the initial condition

$$f(0) = 1$$
$$f(1) = 1$$

describe the famous Fibonacci sequence 1, 1, 2, 3, 5, 8, 13, 21, 34, ...

Question 1.1.3

Suppose rabbits gave birth to 2 pairs each month, but starting after 3 months of life. Can you write down the proper recurrence equation and initial condition to describe this? What are the first 6 terms of the sequence?

The Fibonacci sequence has a peculiar history. The rules Fibonacci put forth for rabbit reproduction are pure fiction. Whatever it was about this that captured the imagination of folks in the 13th century, realism was not it! In the intervening centuries, the Fibonacci sequence has been found to be connected to a surprising number of areas of science and mathematics. See Coxeter [1961] for an account of how the sequence comes up in connection with sunflowers, pineapples and tree branching. Quite recently, the Fibonacci sequence has turned up in connection with finding the best way to arrange and sort data in a technique called polyphase merging (Purdom and Brown, 1985). And the sequence is the basis of "Elliott waves", which have been used to try to forecast the movement of the stock market (*Newsweek*, JA 19, 1987). It seems peculiar, almost mystical, that this sequence should have so many diverse applications.

Question 1.1.4 A Fractal Curve

Figure 1.1.1 shows the Koch "snowflake" curve and how it is built starting with an equilateral triangle. The Koch curve is the "infinite limit" of all the stages. To get from one stage to another, take each segment in the stage, trisect it, and erect an equilateral triangle on the middle segment, pointing out from the interior of the curve. The final step is to erase the middle segment the new triangle is based upon. Let $f(n)$ be the number of sides of the nth stage of the Koch curve. Can you write down a recurrence equation for $f(n)$ and the initial condition?

This curve is an example of a category of curves called fractals which have aroused wide interest lately (see Gleick, 1987; Mandelbrot, 1977; and Thornburg, 1983). The jagged features of various sizes are often considered to make them good models of some natural objects such as leaves, mountain terrain or snowflakes. These fractals are also related to the idea of chaos which we touch on briefly in **Section 1.3**.

Figure 1.1.1 First three stages in the construction of the Koch snowflake curve.

Section 1.2
Solving Recurrence Equations

A recurrence equation plus initial condition can be thought of as a calculating procedure for computing the terms of a sequence, one at a time, starting from the initial terms. But this can be inconvenient. If we have

$$f(n + 1) = 2f(n) \qquad n = 1, 2, 3, \ldots$$
$$f(0) \quad\;\; = 1$$

and we want $f(100)$, we have to work our way up to it, by first getting $f(1)$ from $f(0)$, then $f(2)$ from $f(1)$, then $f(3)$ from $f(2)$, etc. In this example, we can avoid all that because there is a formula that gives the nth term directly:

$$f(n) = 2^n$$

If instead we had

$$f(n+1) = 2f(n) \qquad n = 1, 2, 3, \ldots$$
$$f(0) \quad\;\; = 3$$

then the formula for the nth term would be

$$f(n) = 3(2^n)$$

Where do these formulas come from? Can we always find a formula for the nth term? How do we know the formula is right? Let's investigate.

Verifying a Solution

If your algebra skills are well-developed, you may be able to guess a formula for the solution of a recurrence equation. But how can you be sure your guess is right? Just substitute the alleged solution into the recurrence equation and see if the equation is satisfied. If it is, then see whether the initial condition holds for the alleged solution. If it holds, then you have a solution to the recurrence equation plus initial condition.

Example

Verify that f(n) = $2^{n+2} - 3$ *is a solution to the recurrence equation plus initial condition*

f(n) = 2f(n - 1) + 3   n = 1, 2, 3, . . .
f(0) = 1

To compute 2f(n - 1) + 3, we need first to determine f(n - 1) . This is obtained by substituting n - 1 for n in the formula for f(n). This yields f(n - 1) = $2^{(n-1)+2} - 3$. Now we need to check whether f(n) = 2f(n - 1) + 3 after the substitution is made. We need the following to be true:

$$2^{n+2} - 3 = 2(2^{(n-1)+2} - 3) + 3$$

This equality is easily demonstrated, so we have shown that our formula satisfies the recurrence equation. Next we need to see if the initial condition is satisfied. This means we substitute 0 into the formula and see if it evaluates to 1.

f(0) = $2^{0+2} - 3 = 4 - 3 = 1$

This shows that the given formula satisfies the given recurrence equation plus initial condition.

Question 1.2.1

Can you verify that $f(n) = (1.6)3^n$ is the solution of

$f(n) = 3f(n - 1)$ n = 1, 2, 3, . . .
$f(0) = 1.6$

Question 1.2.2

Can you verify that $f(n) = (1.5)5^n - .5$ is the solution of

$f(n) = 5f(n - 1) + 2$ n = 1, 2, 3, . . .
$f(0) = 1$

Question 1.2.3

Can you verify that $f(n) = (1/\sqrt{5})(\phi^n + (1 - \phi)^n)$ is a solution of
$f(n) = f(n - 1) + f(n - 2)$ n = 2, 3, 4, . . .
$f(0) = 1, f(1) = 1$

where $\phi = (1 + \sqrt{5})/2$? (Note that there are two initial conditions to verify.) Since the given recurrence equation is the one for the Fibonacci sequence, the formula we are dealing with here is a formula for the nth term of the Fibonacci sequence.

Deriving a Solution by Iterated Substitution

It is not always possible to find a formula to describe the solution of a recurrence equation. However, there is a method of finding formulas which works frequently enough that it deserves study. The idea is to do repeated substitutions and then look for patterns in the resulting algebra. (In mathematics, repetition is often called iteration, which accounts for the name of this method.) Let's start by illustrating this for the recurrence equation and initial condition

$$f(n) = f(n - 1) + k, \qquad n = 1, 2, 3, \ldots$$
$$f(0) = a$$

Here k and a are any fixed constants. For example, we might have $k = 2.1$ and $a = -3$. By carrying out our analysis using the symbols k and a we are, in effect, solving an infinite family of recurrence equations.

The recurrence equation can be regarded as a series of equations, one for $n = 1$, another for $n = 2$, etc. We write them out and repeatedly substitute one equation into the next.

$$f(1) = f(0) + k$$
$$f(2) = f(1) + k = (f(0) + k) + k = f(0) + 2k$$
$$f(3) = f(2) + k = (f(0) + 2k) + k = f(0) + 3k$$

If we continue this, it seems evident we should get

$$f(n) = f(0) + nk$$
$$\quad\;\; = a + nk$$

It is easy to verify that this is correct by substitution.

Question 1.2.4

Can you derive the formula $f(n) = ar^n$ for the solution of
$$f(n) = rf(n - 1) \qquad n = 1, 2, 3, \ldots$$
$$f(0) = a$$

Some useful formulas

In solving recurrence equations by iteration we often find it useful to use the formulas for the sums of arithmetic or geometric series. In case you do not remember those formulas, we briefly review them here.

We need a formula for the sum, A, of the first n terms of the arithmetic series starting at 0 and having a common difference of d. That is,

$$A = 0 + d + 2d + 3d + \ldots + (n - 2)d + (n - 1)d$$

We can write the right hand side of the equation backwards, giving

$$A = (n - 1)d + (n - 2)d \ldots + 3d + 2d + d + 0$$

And now adding the two equations together we get

$$2A = (n - 1)d + (n - 1)d + \ldots + (n - 1)d$$

We note that there are n identical terms on the right hand side, so

$$A = n(n - 1)d/2$$

Question 1.2.5

Use this formula to show that the sum of the first n positive integers is $n(n + 1)/2$.

Question 1.2.6

Can you show that the formula for the first n terms of an arithmetic series starting at the value a and having a common difference d is $n(2a + (n - 1)d)/2$?

We will also want a formula for the sum, G, of the first n terms of the geometric series starting at 1 and having a common ratio of r:

$$G = 1 + r + r^2 + r^3 + \ldots + r^{n-1}$$

This time we proceed by multiplying both sides by r and adding 1.

$$rG + 1 = 1 + r + r^2 + r^3 + \ldots + r^{n-1} + r^n$$

We note that the right hand side is G plus r^n, so we make that substitution obtaining

$$rG + 1 = G + r^n$$

and solve the simplified equation for G, under the condition that $r \neq 1$ reaching

$$G = (r^n - 1)/(r - 1)$$

Question 1.2.7

The general geometric series is of the form $a + ar + ar^2 + ar^3 + \ldots + ar^{n-1}$. Can you show that the formula for this sum is $a(r^n - 1)/(r - 1)$ when $r \neq 1$?

In **Section 5.2** we look at another way to prove these summation formulas.

Solving other recurrence equations by iteration

Here are some further examples showing how to solve recurrence equations by iteration.

Example

When we discuss efficiency of algorithms for sorting in Chapter Five, we will find ourselves wanting to know the solution of

$$f(n) = f(n - 1) + n + 1 \qquad n = 2, 3, \ldots$$
$$f(1) = 0$$

Here's how to obtain the solution iteratively.
$f(2) = f(1) + 3 = 3$
$f(3) = f(2) + 4 = 3 + 4$
$f(4) = f(3) + 5 = 3 + 4 + 5$

Evidently, the general pattern is

$f(n) = 3 + 4 + 5 + \ldots + (n + 1)$

This sum is an arithmetic series, and we can use the formula for the sum of an arithmetic series (having n - 1 terms, starting at 3, common difference 1) to obtain

$f(n) = (n - 1)(n + 4)/2$
$ = (n^2 + 3n - 4)/2$

Example

Given this recurrence equation plus initial condition:

$f(n) = rf(n - 1) + k, \qquad n = 1, 2, 3, \ldots$
$f(0) = a$

we carry out iterated substitution as follows:

$f(1) = rf(0) + k = ra + k$
$f(2) = rf(1) + k = r(ra + k) + k = r^2a + rk + k = r^2a + k(r + 1)$
$f(3) = rf(2) + k = r^3a + k(r^2 + r + 1)$
$f(4) = rf(3) + k = r^4a + k(r^3 + r^2 + r + 1)$

It seems that the pattern is

$f(n) = r^na + k(r^{n-1} + r^{n-2} + \ldots + 1)$

The terms in parentheses form a geometric series written in descending order, so we can use the formula for geometric series to replace it and obtain

$f(n) = r^na + k(r^n - 1)/(r - 1)$

Try substituting this formula into the recurrence equation and the initial condition to see if they are satisfied.

Question 1.2.8

In an earlier example we had, $f(n) = 2f(n - 1) + 3$, $f(0) = 1$. This is really a special case of the more general example we just solved. Can you demonstrate that this is so?

So far we have dealt with recurrence equations for which the function f had a value for every non-negative integer input. This is not always the case. In the following example we deal with a function defined for powers of 2.

Example

When we discuss efficiency of computer programs for sorting in Section 5.3, we will find ourselves wanting to solve

$$f(2^{m+1}) = 2f(2^m) + 2^{m+2} \qquad \text{for m = 1, 2, 3, ...}$$
$$f(2^0) \quad = 0$$

Here's how to derive a solution by iteration:

$$f(2^1) = 2f(2^0) + 2^2$$
$$= 2^2$$

$$f(2^2) = 2f(2^1) + 2^3$$
$$= 2\,(2^2) + 2^3$$
$$= 2\,(2^3)$$

$$f(2^3) = 2f(2^2) + 2^4$$
$$= 2\,(2^4) + 2^4$$
$$= 3\,(2^4)$$
$$f(2^4) = 2f(2^3) + 2^5$$
$$= 3\,(2^5) + 2^5$$
$$= 4\,(2^5)$$

Evidently, the pattern is

$$f(2^m) = m\,2^{(m+1)}$$

Try verifying this solution by substitution.

Question 1.2.9

Try using the method of the previous example to solve:

$$f(2^{m+1}) = 2f(2^m) + 2m + 1 \qquad m = 1, 2, 3, \ldots$$
$$f(2^0) = 0$$

So far, our examples of iterated substitution can all be regarded as based on the following way of thinking:

"I know $f(0)$.
Using $f(0)$ I can find the value of $f(1)$.
Using the value of $f(1)$ I can find the value of $f(2)$.
By carrying this out n times I can get $f(n)$."

Another form of iterated substitution is based on the reverse way of thinking. It could be summarized like this:

"I could get $f(n)$ if I knew $f(n-1)$.
I could get $f(n-1)$ (and, therefore, $f(n)$) if I knew $f(n-2)$.
I could get $f(n-2)$ (and, therefore, $f(n)$) if I knew $f(n-3)$.
By carrying this logic often enough, I can get $f(n)$ from the initial condition."

Here is an example of this "reverse" form of iterated substitution.

Example

Let's examine again the equation we used to begin our discussion of iterated substitution

$$f(n) = f(n-1) + k, \quad n = 1, 2, 3, \ldots$$
$$f(0) = a$$

By substituting $n-1$ *for* n *in the recurrence equation we obtain*

$$f(n-1) = f(n-2) + k$$

Substituting this into the original recurrence equation gives

$$f(n) = (\, f(n-2) + k \,) + k = f(n-2) + 2k$$

Now substitute n - 2 *for* n *into the original recurrence equation to obtain*

f(n - 2) = f(n - 3) + k

Substituting this into the previous equation gives

f(n) = f(n - 3) + 3k

If we were to continue this for a total of n - 1 *steps, we would arrive at*

f(n) = f(0) + nk

Since we know that f(0) = a, *we have our solution:*

f(n) = a + nk

Reverse iterated substitution is closer than the forward form to the way of thinking which we must adopt in order to deal with recursive algorithms in Chapter Three.

Question 1.2.10

Use reverse iterated substitution to find the formula for

$$f(n) = rf(n - 1), \qquad n = 1, 2, 3, \ldots$$
$$f(0) = a$$

Question 1.2.11

Use reverse iterated substitution to find the formula for

$$f(n) = f(n - 1) + n + 1, \qquad n = 2, 3, 4, \ldots,$$
$$f(1) = 0$$

Question 1.2.12

Use reverse iterated substitution on the recurrence equation

$$f(n) = rf(n\text{-}1) + k, \qquad n = 1, 2, 3, \ldots,$$
$$f(0) = a$$

Before we leave the subject of formulas for recurrence equations, we should note that not every recurrence equation has a solution in the form of a formula involving the familiar elementary functions (such as polynomials, exponentials, logs, and trigonometric functions). An important example is

$$f(n) = c\,(\,1 - f(n-1)\,)\,(\,f(n-1)\,), \qquad n = 1, 2, 3, \ldots,$$
$$f(0) = a$$

This equation is not just a curiosity. As we shall see in the next section, it describes a population undergoing "limited growth" and is of great interest to biologists.

Some terminology

We have been thinking of $f(n)$ as a sequence of numbers. Another way to think of a sequence is as a special kind of mathematical function. A sequence is a function whose domain is the positive, or maybe non-negative, integers. Finding the nth term of a sequence is equivalent to evaluating the function at the point n.

We have dealt with two methods of defining a function (sequence) in this chapter. One method is to have a formula for the function, such as $f(n) = 2^n$. The other method is to define the function by stating a recurrence equation plus initial condition which the function satisfies. In the examples where the recurrence equation plus initial condition allowed us to find an equivalent formula definition, we feel confident that we had the "right" answer. For those cases, such as limited growth, in which a formula definition is not obtainable, we can believe in the existence of and uniqueness of a function behaving in the designated manner because we know that starting with the initial condition and applying the recurrence equation as often as needed, we can get exactly one answer for the function value at any chosen value of n. So even if we don't have a formula, we do have a method by which to evaluate the function; so it really is a function – it just looks different to us.

Section 1.3
Population Growth

Suppose a colony of yeast reproduces by this rule: Every hour, 36% of the yeast split in two. We also assume that yeast never die and there is no other form of reproduction. These assumptions are very realistic for some circumstances. Later we'll examine a situation where a different set of assumptions is in order. Let $f(n)$ denote the number of yeast at n hours after the start of observation. Our assumptions imply

$$f(n) = f(n - 1) + .36f(n - 1)$$
$$= 1.36f(n - 1) \qquad n = 1, 2, 3, \ldots$$

In **Section 1.2** we saw how to find a formula for the solution of this recurrence equation, namely

$$f(n) = (1.36)^n f(0)$$

For example, suppose there were 96 yeast to start with. Then **Figure 1.3.1** shows a plot of this function for $n = 0, 1, 2, 3, 4$. The formula on the right side of the above equation is an example of an exponential function because the variable n is in the exponent. For this reason, the yeast in this example are said to be undergoing exponential growth.

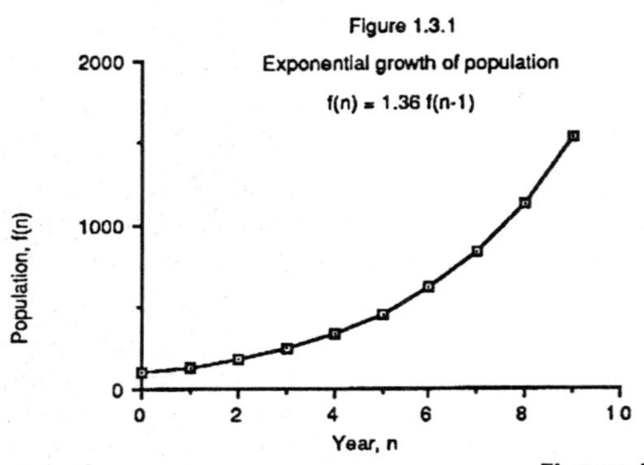

Figure 1.3.1

Question 1.3.1

How many hours will it take for the yeast population to double from its original size of 96 yeast? How long will it take for another doubling? Let t^* be the time it takes to double from 96 to 192.

Can you show that, no matter what starting time t we choose, after t^* additional time units, the population is twice the size it was at t?

Question 1.3.2

Here's how one can prove that an exponential function grows beyond all bounds. Let c be any number >1 (like 1.36 in our population example) and M any other positive number. To show that $c^n > M$ when n gets large enough, begin by setting $c = 1 + e$ where $e > 0$ and then expand $(1 + e)^n$ by the binomial theorem (Niven, 1965). Can you see how to finish the proof?

Exponential decay

Carbon-14 is an isotope of carbon which undergoes radioactive decay into carbon-12 according to the rule that in each thousand years, the fraction of an initial amount of carbon-14 which transforms itself is .114.

For example, if there were 500 grams of carbon-14 in an object in the year 990 AD, then by 1990 AD, 1000 years later, $(500)(.114) = 57$ grams would be transformed, leaving 443 grams of carbon-14. After another 1000 years $(443)(.114) = 50.502$ additional grams of carbon-14 will have turned to carbon-12.

Question 1.3.3

Let $f(n)$ be the fraction of the original amount of carbon-14 which remains after n thousands of years. When $n = 0$, the entire amount is present, so the fraction of the original amount present is 1. Thus, $f(0) = 1.0$. What is the relation between $f(n)$ and $f(n - 1)$? What is the solution to this recurrence equation? How is this solution similar to the one we got for exponential growth? This similarity is why we call radioactive decay by the name exponential decay. See Cozzens and Porter [1989] for applications of this idea to archeological dating, for example, of the alleged remains of Noah's ark.

The reality of limited growth

Figure 1.3.1 suggests that the yeast cells will grow without limit. Indeed, it can be proved that the function $96(1.36)^n$ grows without bounds. (see **Question 1.3.2**) No matter how large a number you choose – whether it be 10^5, 10^{10}, 10^{100} or anything – there will be a corresponding value of n which makes $96(1.36)^n$ exceed the number you chose. In other words, however large you want the yeast colony to be, it will get to that size if you wait long enough.

Needless to say, this can't be really true. Otherwise the world would long ago have been overrun by yeast. While a small yeast colony with plenty of food will start out growing according to our assumptions, eventually some slowing down of the growth will have to occur. The rate of splitting must slow down or some yeast must die or both.

This relatively simple line of mathematical reasoning about exponential growth is responsible for one of the most important debates that has ever occurred in our political and intellectual history. The story starts with Thomas Malthus, an English country parson who asserted that human population has a tendency to grow according to an exponential law just as yeast does. Since the earth's resources are finite, population can't grow toward infinity indefinitely. Therefore, something must occur to frustrate the tendency to exponential growth. There are only two possibilities: birth rates will have to decline or we'll have to be prepared for death rates to rise.

There isn't really any controversy about this part of Malthus' argument. The argument arises when economists and politicians try to decide whether we are close to the overpopulation point now and whether drastic measures – such as vigorous birth control measures or limits to growth need to be instituted right now. In the last 200 years there has been an ongoing debate between "prophets of doom" and "apostles of growth." The prophets of doom typically claim that we are near the point of famine or ecological disaster. The apostles of growth claim that human ingenuity is still capable of growing more food, preserving the crucial parts of the ecology and generally troubleshooting any problems that arise for quite some time yet (Cole, et al, 1973 and Meadows, et al, 1972).

It seems likely that both camps have captured some of the truth. There have been famines, ecological problems and occasional resource shortages, but only in some parts of the world. In general, the human race is more numerous than ever, and much of it is healthier and more prosperous than it was a few decades ago – all of which suggests that up until now the optimists have won the argument. But there is always the future to think about. Since the argument is really about the future, and the future has not arrived, the pessimists can never be completely ignored.

Recurrence equations for limited growth

Biologists have developed recurrence equations which predict a population changing in a way that is more realistic than indefinite exponential growth. These equations describe limited growth. When a population is small in relation to the space, food and other resources available to it, it seems reasonable that it would grow – just as our yeast did in the previous discussion – according to an equation of the form

$$f(n) = c\,f(n-1) + f(n-1) \qquad\qquad n = 1, 2, 3, \ldots$$

This gives rise to the solution $f(n) = (1 + c)^n\, f(0)$. The constant c reflects the difference of births and deaths. When $c > 1$ (the typical Malthusian scenario) exponential growth results.

However, if the resources are fixed, as the size of the population increases, the potential for growth becomes limited and the growth rate should decline. One way to model this is to use a growth rate which is not a constant c but which is a declining function of the population size. Suppose, for example, that the maximum population that could be supported by the resources at hand is K. Then it would be reasonable to have a growth rate which is approximately c when the population is small but which declines to 0 as population increases toward K. One growth rate (among many) which has this character is

growth rate = $c(1 - \text{population size}/K)$.

Using this in place of the constant c in the recurrence equation we started with gives this equation:

$$f(n) = c(1 - f(n-1)/K)\,f(n-1) + f(n-1) \qquad\qquad n = 1, 2, 3, \ldots$$

Figure 1.3.2 shows how a population governed by this law would change over time.

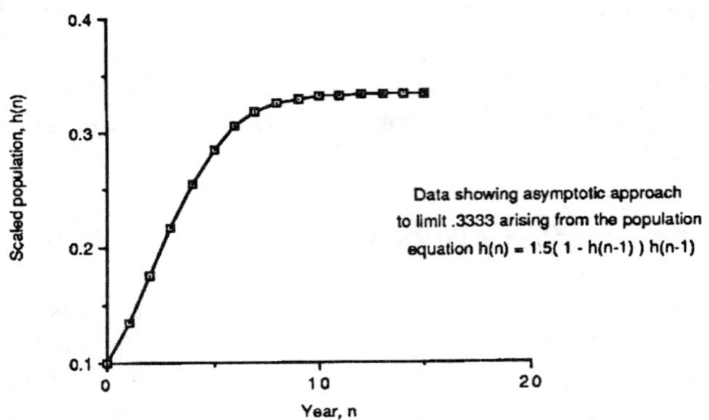

Figure 1.3.2

Question 1.3.4

The above form of the recurrence equation for limited growth can be simplified algebraically to this:

$$h(n) = b(1 - h(n - 1))h(n - 1)$$
$$= b*h\,(n - 1)\,(1 - h(n - 1))$$

This is the form alluded to at the end of **Section 1.2** and which will be used in later sections. Here are some hints for how to do this simplification:

1. Factor $f(n\text{-}1)$ in the right hand side of the original; divide the equation by K; make the substitution $g(n) = f(n)/K$. You should get

$$g(n) = (c(1 - g(n - 1)) + 1)\,g(n - 1).$$

2. Multiply by the c; then multiply both sides of the equation by an additional factor of c; make two substitutions: $b = c + 1$ and $h(n) = (c/(c + 1))g(n)$. You should arrive at the predicted form.

Asymptotic Limits and Chaos

We have remarked at the end of **Section 1.2** that there is no known elementary formula for the solutions to the recurrence equation :

$$h(n) = b * h(n - 1) (1 - h(n - 1))$$

which we showed in this section to be related to limited population growth (see **Question 1.3.4**). One might suppose that this would rule out learning much about the solutions to that equation. However, this is not true at all. A good deal has been discovered about the "asymptotic behavior" of its solutions. By asymptotic behavior we mean what happens to $h(n)$ as n increases toward infinity.

Figure 1.3.2 shows one situation which might occur. The values of $h(n)$ get ever closer to some particular limit value L, which in **Figure 1.3.2** seems to be approximately .3333. This asymptotic behavior is called approaching a limit.

In **Figure 1.3.2** we used the limited growth equation with $b = 1.5$ and an initial condition of $h(0) = 0.1$. One might ask whether, if we change the values of b and $h(0)$, we always get the same kind of asymptotic behavior. The answer is definitely not. **Figure 1.3.3** shows what can occur when $b = 3.1$: as n increases. The values of $h(n)$ tend toward a regular periodic alteration among fixed values. This asymptotic behavior is called asymptotic periodicity. The number of values among which $h(n)$ tends to alternate is called the period. In **Figure 1.3.3** the period is 2.

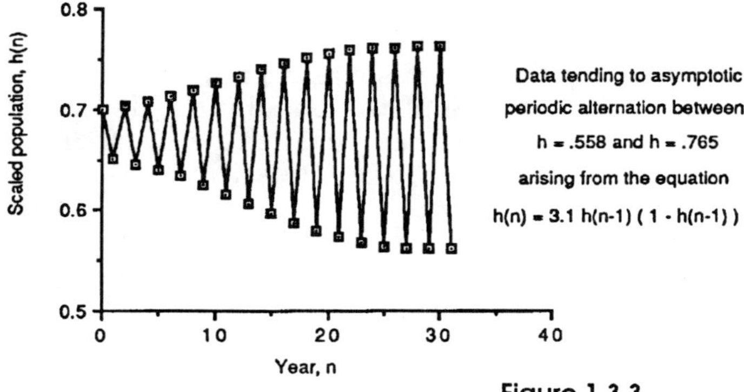

Figure 1.3.3.

Periodicity and approaching a limit are asymptotic behaviors which mathematicians have studied for hundreds of years and which hold few surprises today. But there is a third kind of asymptotic behavior our limited growth recurrence equation can generate which has been investigated extensively only in the last few decades and which is raising quite a few eyebrows. This behavior is called chaos and is illustrated in **Figure 1.3.4** where we have taken b = 3.75 and h(0) = 0.5. Here the successive values of h(n) seem to change randomly with no particular pattern. In fact it almost seems as if successive values of h(n) are determined by some chance device such as throwing a dart at a number line. What scientists have found remarkable about this is that while the appearance is of randomness, the equation which generates this data has nothing probabilistic or even very complicated about it.

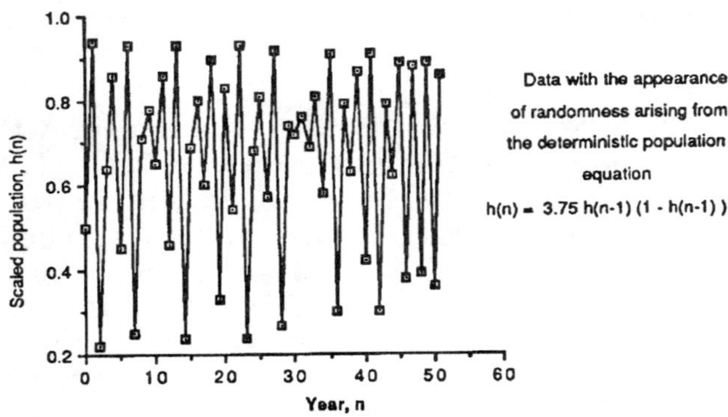

Figure 1.3.4

To appreciate what this means, suppose we were studying the population of some organism in an ecosystem. and we tabulate the size of the population, $f(n)$ at the end of the nth time period. We then scale these numbers using the changes of variable described in **Question 1.3.4** and obtain the series of numbers, $h(0)$, $h(1)$, $h(2)$, . . ., which have a chaotic appearance to them as in **Figure 1.3.4**. How could we explain such chaotic population fluctuations? The following have been some of the more traditional approaches:

1. The "Nature Is Random" explanation. This explanation asserts that nature is inherently random and some parts of it cannot be described by equations that allow prediction of what will happen next.

2. The "Nature Is Too Complicated" explanation. This explanation asserts that, although nature is not random, there are so many influences on the population size of the organism – weather, competing species, food supply, effect of man on the environment, etc. – that the combined effect of all these factors makes things look random.

However, **Figure 1.3.4** shows that we don't need to accept either of these conclusions. The limited growth equation is entirely deterministic (contains no chance factors) and is extremely simple, but it can generate chaotic data.

Population biology is not the only area where simple equations can generate chaotic behavior. One of the first areas where chaos was discovered is meteorology – the study of weather and climate. Subsequent examples have been found in various branches of physics as well as other disciplines. The study of these matters has exploded recently and has suggested to some scientists that science has reached a significant turning point in its development. Even more cautious scientists find the ideas highly intriguing.

A popular account of the wide spectrum of chaos throughout the sciences (with no equations but nice pictures and analogies) may be found in Gleick [1987]. A more mathematical approach that focuses particularly on the limited growth equation can be found in Hofstadter [1981]. For a mathematical treatment of the other two types of asymptotic behavior for the limited growth recurrence equation, approach to a limit, and periodicity, see Chapter 5, Section 2 and especially Subsection 2F of Meyer [1984]. See Devaney [1989] for an count of chaos especially written for high school students.

Chapter Two
Recursive Definitions

I N CHAPTER ONE WE DEFINED FUNCTIONS USING RECURRENCE equations. In a way, we can view a recurrence equation as a definition of a function "in terms of itself". Determining the value of the function at a number n is based on knowing the value of that same function at values less than n. A definition which behaves in this self-referential manner is said to be a recursive definition. The word recursion is used to describe this technique of self-reference. Thus we say that a recursive definition uses recursion. In this chapter we will study recursion as a means of defining linguistic and algebraic expressions instead of functions.

Section 2.1
Recursion and Language

When linguists attempt to write down the grammar rules for a natural language such as German, English or pidgin English, they often find themselves using recursion in their descriptions. In this section we give a brief introduction to syntax diagrams, which are methods for describing grammar rules, and the use of recursive syntax diagrams.

A typical sentence structure used in English is a noun phrase followed by a verb phrase. This structure is described symbolically with what is called a syntax diagram (see **Figure 2.1.1**).

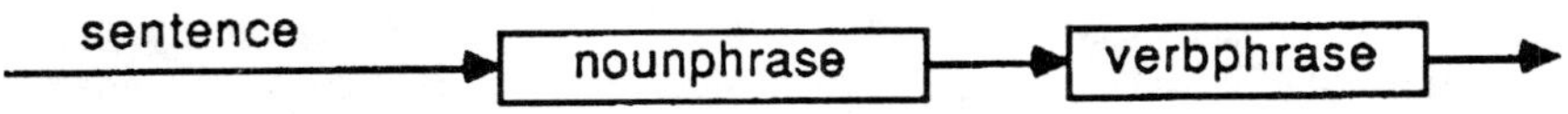

Figure 2.1.1

Think of this diagram as a guide to building a sentence. We begin at the left (where the linguistic entity to be built is described) and then follow the arrows, writing down an example of whatever is called for in a particular box when we encounter that box. An example is shown by the words in italics in the syntax diagram in **Figure 2.1.2**.

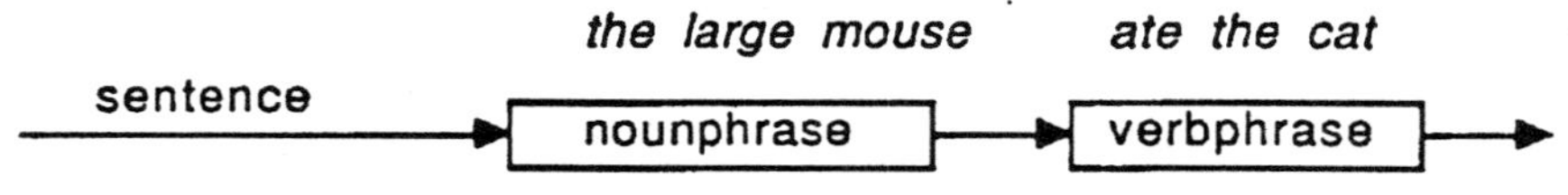

Figure 2.1.2

A string of words that doesn't conform to this structure is "eat your beans" because this sentence starts with the verb phrase "eat" instead of starting with a noun phrase as required by **Figure 2.1.1**. Since "eat your beans" is actually a sentence, this example shows that one needs many syntax diagrams to describe any particular linguistic structure, such as a sentence in English. A string of words qualifies as a sentence if it satisfies any of these syntax diagrams. For simplicity, we consider only sentences satisfying the syntax diagram in **Figure 2.1.1**.

Noun phrases themselves come in a variety of structures. **Figure 2.1.3** shows some noun phrases, together with the structures they exemplify.

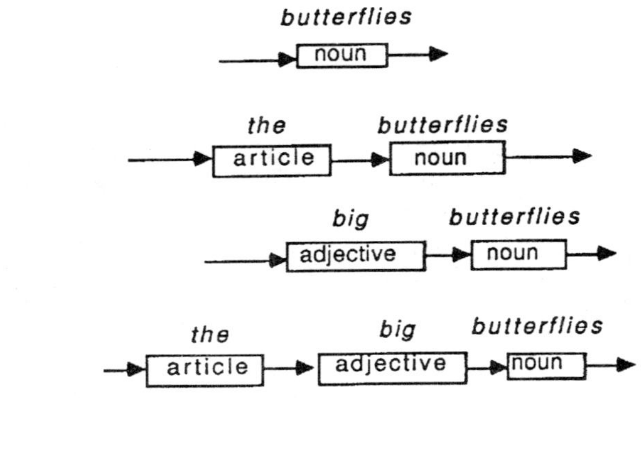

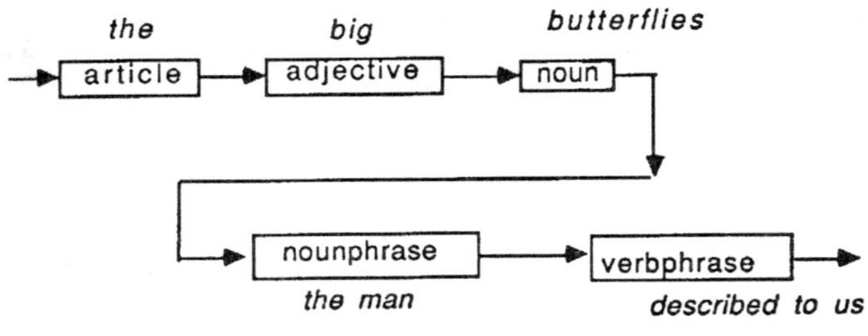

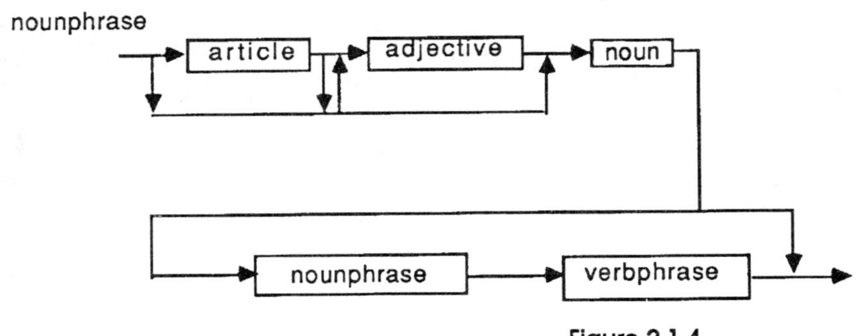

Figure 2.1.3

All of the five types of noun phrases shown in **Figure 2.1.3** can be summed up in one syntax diagram, as shown in **Figure 2.1.4**.

Figure 2.1.4

In following the diagram in **Figure 2.1.4** to build a noun phrase, there are various points where choices can be made. For example, we can bypass article and adjective and build our noun phrase starting with a noun. Or we can start with an article and bypass the adjective. All the noun phrases we have seen fit the complex syntax diagram in **Figure 2.1.4**.

When we begin to build a noun phrase using the syntax diagram of **Figure 2.1.4**, on our first pass through we cannot use the branch that goes through the box labeled "noun phrase" because no noun phrase has been built yet. We would have to start by using a route through the diagram that does not encounter this box. This is a general rule about using syntax diagrams to build entities in a language: We can take any route through the diagram provided we have something available to "fill in the blank" each time we encounter a box.

But the most interesting thing about **Figure 2.1.4** is not the system of bypasses, but the use of "noun phrase" in the middle of a diagram that is meant to describe what a noun phrase is. This is an example of recursion. For example, let's see how the sentence "the big butterflies the man described to us" can be built using this recursive definition. To begin with, we first build the noun phrase "the man." To do this, start at the left of the syntax diagram and write down the article "the." Next, bypass the adjective box and write down the noun "man." Finally, bypass the last two boxes ("noun phrase" followed by "verb phrase") and quit. Now let's go through the diagram again to build the noun phrase "the big butterflies the man described." We begin at the left and write down the article "the," followed by the adjective "big". This brings us to the box labeled "noun" and for this we write down "butterflies." Now instead of bypassing the "noun phrase" and "verb phrase" boxes, we'll use the noun phrase already built, "the man." Finally we'll write down the verb phrase "described to us" and quit. (We won't describe syntax diagrams for verb phrases, so let's take it on faith that "described to us" is a proper verb phrase.) This example shows that if we are building a noun phrase using this syntax diagram, we must use the bypass around the last two boxes unless we already have a noun phrase (and verb phrase) previously built. This means that somewhere we must have gone through the diagram using the bypass of the last two boxes before we could build a noun phrase using the last two boxes. This is akin to the fact that to solve a recurrence equation one must have an initial condition.

Our recursive definition of noun phrase can be used to describe even more complex sentences with many levels of noun phrases within noun phrases. Consider for example the sentence:

the big butterflies the man John met described to us

This is closely related to the one we just analyzed: the difference is that in place of the simple noun phrase "the man," we have a noun phrase, "the man John met" which contains a noun phrase, "John," within it. To build "the man John met," we first need to build the noun phrase "John." We do this by going

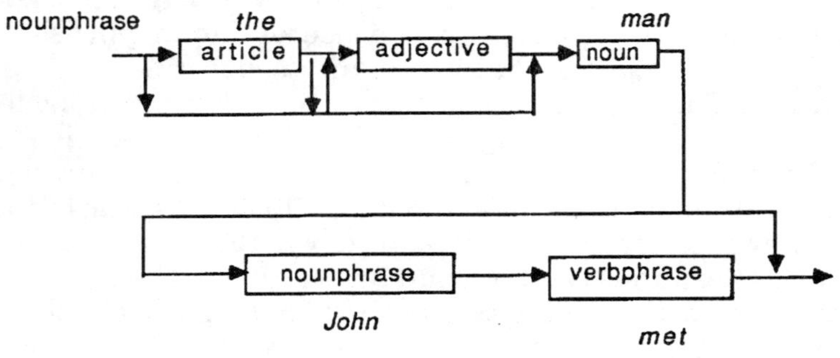

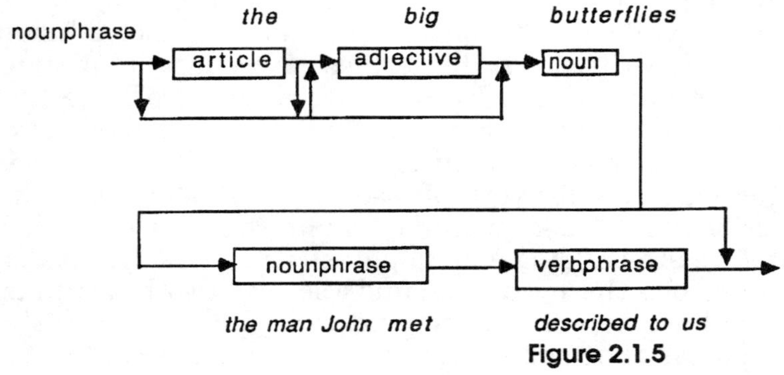

Figure 2.1.5

through the syntax diagram and stopping only at the box labeled "noun" and writing down "John," bypassing all other boxes. Next we build the noun phrase "the man John met" by going through the syntax diagram twice from left to right and responding to the boxes as shown in **Figure 2.1.5**.

Question 2.1.1

Can you develop a syntax diagram for the name of a person using such elements as title, first name, middle name, last name, initial, . . . ? Can you develop the supporting syntax diagrams for the elements?

Question 2.1.2

If you have ever seen flowcharts used to describe algorithms, then you can think about the similarities and differences between flowcharts and syntax diagrams. For starters, both contain directed paths – arrows. What else do they share?

Question 2.1.3

Here is a syntax diagram for a gortimaldinger:

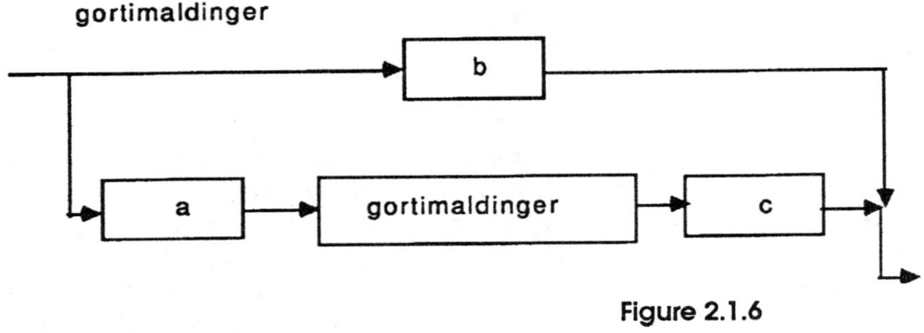

Figure 2.1.6

a. Which of these are examples of gortimaldingers:

b, abb, abc, abbc, aaabccc, aabccc, a, c, ac.

b. Try to give a verbal description, without drawing a syntax diagram, of what it takes for a character string to qualify as a gortimaldinger.

Question 2.1.4

Construct a syntax diagram which describes the set of all finite strings containing just two symbols, a and b, and where the string starts with a certain number of a's and ends with an equal number of b's.

Section 2.2
Syntax Diagrams for Algebraic Expressions

Syntax diagrams are used not only to describe natural language structures but also to describe structures in computer languages such as BASIC or Pascal. These languages, like all computer languages, and like mathematics itself, have rules for what well-formed algebraic expressions are.

Question 2.2.1

Find a Pascal text that contains syntax diagrams for the Pascal language (many Pascal texts do contain them). Do these diagrams help you write syntactically correct Pascal? Are any of the diagrams recursive? Do these diagrams help you write a Pascal program that does the job correctly, assuming that the Pascal is written correctly?

When you write a computer program that involves an algebraic expression such as $a - (b + (c - d))$, some piece of software (usually a compiler or interpreter) needs to figure out what this means and whether it even makes sense in the language you are writing in. For example, if you leave out a parenthesis and write $a - (b + (c - d)$, the compiler or interpreter should present you with an error message of some kind. In our example, the error was easy to catch: The right and left parentheses are unequal in number in the incorrect version of the expression. But here is an incorrect expression where the right and left parentheses are equal in number, but something else is wrong:

$$a + (b - c) + d) - e + (x + (-z)$$

Question 2.2.2

What principle could we use to reject this string of symbols?

In addition to problems with parentheses, there may be problems with the operators +, -, *, / as in the following incorrect expressions:

$$a + * b$$

$$a + (*b + c)$$

It might be possible to make a complete list of errors that make the syntax of an algebraic expression incorrect. However, the approach usually taken by computer scientists is to set out one or more syntax diagrams which describe correct algebraic expressions. This set of one or more syntax diagrams is called a "recognizer" for algebraic expressions. **Figure 2.2.1** shows a diagram that can serve as a recognizer for a limited class of algebraic expressions. (We are not going to deal with exponents in this simple example.) Notice that this syntax diagram is recursive because the middle and lower branches involve the concept of "algebraic expression."

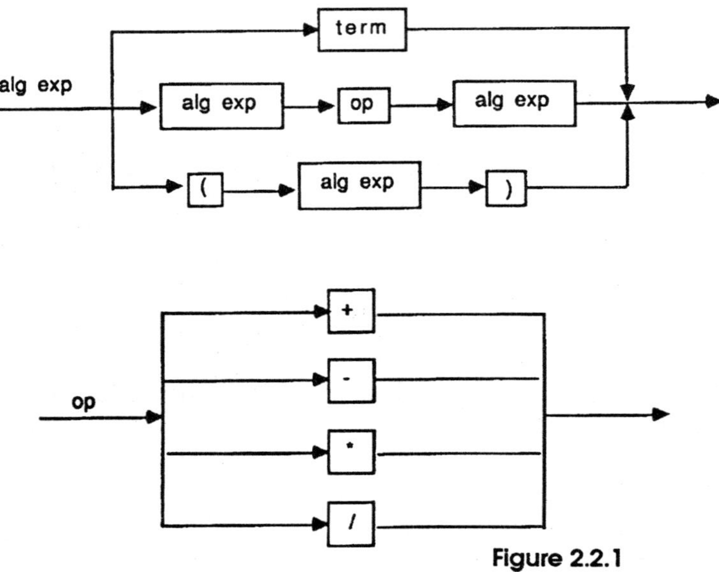

Figure 2.2.1

In the top syntax diagram, "term" means any numerical value such as 4, 5.7, or -.00005, or any variable name standing for a number, such as x, y or temp. The upper branch of the top syntax diagram tells us that if we simply write down a variable name or numerical value we have an algebraic expression.

Question 2.2.3

Can you construct a syntax diagram for a "term"? The basic elements should include digits, decimal point, letters (for variable names), plus and minus signs.

Question 2.2.4

Can you construct a syntax diagram for positive integers written with commas to separate groupings of three digits, for example 12,500 as opposed to 12500?

The middle path through the top syntax diagram of **Figure 2.2.1** can be used to connect two previously built algebraic expressions with +, -, *, /. For example, since 3, 4, x, and y are algebraic expressions, then so are:

$$3 + x$$
$$3 - x$$
$$3 * x$$
$$3/x$$
$$3 + 4$$
$$4 - x$$

etc.

By applying this middle path more than once we can determine that the following are algebraic expressions:

$$3 + x + y$$
$$3 - 4/x + 2$$
$$4/y + 7 * 5 - x$$

etc.

By also using the last branch of the top syntax diagram we can get the following expressions:

$$(4 + x)$$
$$(4/x)$$
$$(x - 3)$$
$$(2 + y - x) * (x - 8)$$

etc.

If we have a complicated algebraic expression we can build it "from the inside out" using the syntax diagrams of **Figure 2.2.1**. For example,

$$((1.5 - x) * (x + y / (7 + z)) + x) - 1$$

The inside-most expression is $7 + z$ and we build this first. We start by noting that since 7 and z are terms, they are also algebraic expressions (according to the upper branch of the top diagram). Using the syntax diagram's middle branch, we get that $7 + z$ is an algebraic expression.

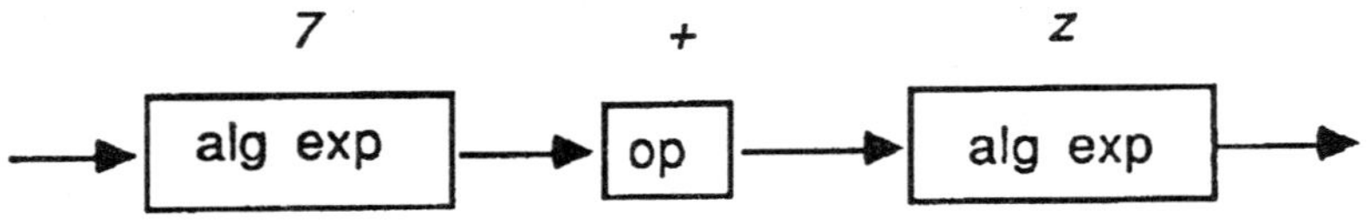

Next use the lower branch as follows

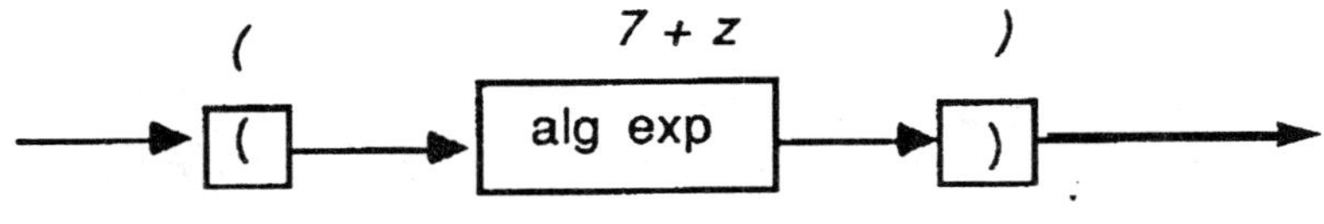

to obtain that $(7 + z)$ is an algebraic expression.

Next use the middle branch twice to obtain $x + y / (7 + z)$

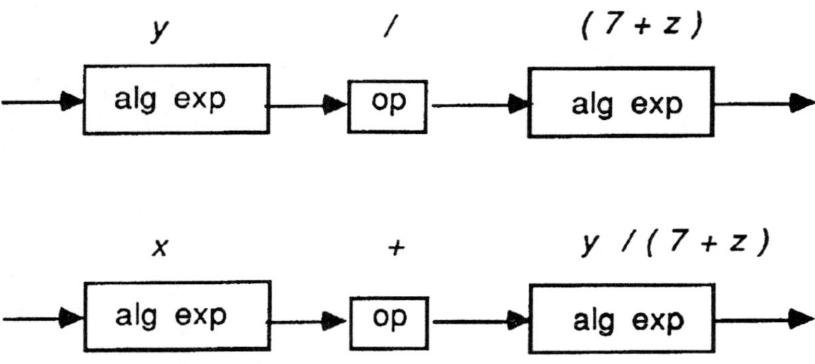

And we continue in this way until we obtain the entire expression. If we are unable to reach the expression by some combination of top, middle and bottom branches, then the expression would not be a (well-formed) algebraic expression.

Question 2.2.5

In logic or boolean algebra there are rules for forming correct expressions, often called well-formed formulas or WFF's. Here are those rules in paragraph form. Can you draw the corresponding syntax diagram?

1. Any capital letter is a WFF.
2. If P and Q are WFF's then the following are also WFF's:
   (P)
   (notP)
   (P and Q)
   (P or Q)
   (P implies Q)
   (P if and only if Q)

Question 2.2.6

Here is a definition of a set K of numbers. Can you draw syntax diagram for "element of K"?

1. 6 is an element of K.
2. If x and y are elements of K then so are $x + y$ and $x + 5$.

Question 2.2.7

Can you develop a test procedure to determine if a given positive integer is an element of the set K as described in **Question 2.2.6**?

Chapter Three
Recursive Algorithms

Section 3.1
Recursive Functions

IN CHAPTER ONE WE STUDIED RECURRENCE EQUATIONS. A recurrence equation implicitly defines a function by expressing a relationship among values of the function evaluated at different integers. We saw that under some circumstances it is not hard to find a formula definition for a function defined by a recurrence equation. If a formula is not available for such a function, it is often possible to compute values of the function using a recursive algorithm expressed in a computer language such as Pascal. In this section, we introduce such algorithms.

Evaluating a recursively defined function

The recurrence equation for limited growth, which we met in **Section 1.3**, is not solvable in terms of a formula involving elementary functions and operations. This equation can be written as

$$f(n) = b*f(n-1)*(1-f(n-1)) \qquad n = 1, 2, 3, \ldots$$

where b is a constant. Suppose that we need to know the value of this function f for some specified value of n. For example suppose $b = 2$ and $f(0) = .25$ and we wish to know $f(3)$, i.e. $n = 3$. We can write the resulting definition of f in this way:

$$f(n) = 2*f(n-1)*(1-f(n-1)) \qquad n = 1, 2, 3, \ldots$$
$$f(0) = .25$$

This definition encompasses both the recurrence equation and the initial condition. Suppose we tried to incorporate this definition into an algorithm for calculating values of f, as in the following pseudocode:

> if $n = 0$
> then $f(n) = .25$
> else $f(n) = 2*f(n - 1)*(1-f(n - 1))$

Is this really an algorithm a computer could execute? It is obvious how to follow this algorithm for finding $f(0)$, but perhaps not so obvious how to use it to evaluate f at higher values of n. In fact at first glance this algorithm may look unworkable because for $n>0$, $f(n)$ is calculated assuming that $f(n - 1)$ is available. How can we guarantee this availability? Our original choice of $n = 3$ will serve as an example.

To find $f(3)$ we first need $f(2)$. In order to get $f(2)$, we must know $f(1)$. And to get $f(1)$, we need $f(0)$. But – and here is what saves us from an unworkable algorithm – $f(0)$ is defined without reference to any other value of f; $f(0) = .25$. Substituting .25 for $f(0)$ we see that $f(1) = .375$; substituting that value into the expression for $f(2)$ we get $f(2) = .46875$. And finally, proceeding to $f(3)$ we get a value of .4980. A computer goes through a process like this in response to a properly written "recursive program" with the structure sketched by our pseudocode.

The algorithm we just used to evaluate the function f is called a recursive algorithm because, for some values of n, $f(n)$ is defined in terms of f evaluated at smaller values of n. Often this is referred to loosely as "f is defined in terms of itself." It is natural to use a recursive algorithm to evaluate a function having a recurrence equation as part of its definition. But recursion is not limited to this application; recursive algorithms are used in many other circumstances.

Evaluating a polynomial

If you have to evaluate a polynomial function such as

$$p(x) = 7x^5 + 82x^4 - 23x^3 - 4x^2 + 51x - 1$$

for $x = 13$, whether or not you use a calculator you want to organize your work efficiently. Many people approach this task by evaluating the terms of the polynomial in ascending order.

They find 51 times 13 and then find $13^2 = 169$. Next, the 169 is used to find both $4x^2$ and the value of x^3. The key point is that 13^3 is evaluated with only one additional multiplication as $13*(169)$ instead of starting from scratch and multiplying three 13's together. The evaluation continues in this way, from right to left, until all terms are done. Although the polynomial was written in descending powers of x, it is "better" to evaluate it as though it were presented in ascending order, because it is more efficient to find x^n from its relationship to x^{n-1} than to find those two values independently and ignore their relationship. In general $x^n = x*x^{n-1}$. If we think of x^n as a function f, then we note that f depends on both x and n; f is a function of two variables, $f(x, n)$. We can write a recursive definition of $f(x, n)$:

$$f(x,n) = x*f(x, n - 1) \qquad\qquad n = 1, 2, 3, \ldots$$
$$f(x,0) = 1$$

And just as in the growth example above, there is a corresponding recursive algorithm for evaluating the function.

Question 3.1.1

Can you construct the algorithm? Use as a pattern the pseudo-code we gave previously for the limited growth example. (*Hint*: The algorithm must tell how to find $f(x, n)$ for any x and for non-negative integer n. The "if statement" will depend only on n.)

Question 3.1.2

Here is another function defined recursively. Do you recognize it?

$$f(n) = n*f(n - 1) \qquad\qquad n = 1, 2, 3, \ldots$$
$$f(0) = 1$$

Question 3.1.3

Do you recognize this function?

$$f(n) = f(n - 1) + f(n - 2) \qquad\qquad n = 2, 3, 4, \ldots$$
$$f(0) = 1$$
$$f(1) = 1$$

Question 3.1.4

Can you construct the corresponding recursive algorithms for evaluating the functions in **Questions 3.1.2** and **3.1.3**? Use as a pattern the pseudocode we gave previously for the limited growth example.

Question 3.1.5

Polynomials such as $p(x)$ are sometimes written in a different algebraic form in computer programs. Can you think of any reasons why it might be preferable to write the polynomial above as

$$p(x) = -1 + x(51 + x(-4 + x(-23 + x(82 + x(7)))))$$

Recursive functions in general

Let us leave the specific examples for a moment and look at the thought process leading to recursive definitions of functions. We start with a function f defined for non-negative integers, n. Upon examination of the function we discover that we could evaluate the function for a particular value of n if we knew the value of the function for some smaller value(s) of n. That is, we know how to put together "previous answers" to get the answer we want; all we need is the previous answers. And we need some insurance: we need to know that somewhere, for some special case(s) of n, there really is an answer that we can get without "previous answers" being involved. In general, we have this definition of the function:

$$f(n) = \text{combination of} \quad \text{"previous answers"} \qquad \text{if } n \text{ is not a special case}$$

$$f(n) = \text{known value} \qquad \text{if } n \text{ is a special case}$$

Since the special case consists of the lowest value(s) of n, reaching it stops the recursive process by giving us a known value, and causes us to exit from the recursion, so it is called the non-recursive exit. Without the non-recursive exit, our function definitions would lead to evaluation algorithms which do not terminate.

Question 3.1.6

Can you identify the non-recursive exits in all the recursively defined functions we have discussed?

Corresponding to the general definition given above we have this general algorithm for evaluating the function:

> **if** n is a special case
>> **then** $f(n)$ = known value
>> **else** $f(n)$ = combination of "previous answers".

In summary we see that a recursive definition of a function of n, and the corresponding evaluation algorithm, must have at least two parts: The general part involves knowing how to combine function values for smaller values of n to get the definition of f for this value of n. The other part is the special case of the non-recursive exit. The essence of recursion is understanding how the general, or recursive, part of the algorithm works. As we shall soon see, many problems besides evaluating a function can be profitably solved by a recursive solution algorithm.

Recursive computer functions

Most high level computer languages have ways of constructing subprograms called functions. A function, in computing as in mathematics, accepts input parameters or independent variables and returns exactly one result or dependent variable. Not all high level languages allow functions to be evaluated by recursive algorithms. Languages allowing recursion include Pascal, Logo, PL1, C, Ada, and some versions of BASIC; languages not accommodating recursion include COBOL, RPG, and most versions of FORTRAN. Newer computer languages are usually written to support recursion and older ones are not. We now assume that you have available some language which does allow recursion; our examples are in Pascal. In **Section 5.1** we discuss how to simulate recursion in those languages not accommodating it.

Here is our general recursive evaluation algorithm written as a Pascal-like function:

```
function F(n: 0 .. maxint): 0 .. maxint;

begin
   if n is a special case
     then
        F(n) : = known value
     else
        F(n) : = combination of "previous answers"
 end;
```

And here is a Pascal function to evaluate the factorial function, $f(n) = n!$, from **Question 3.1.2** (do you recognize it?):

```
function FACTORIAL(n: 0 .. maxint): 0 .. maxint;


begin
  if n = 0
    then
       FACTORIAL : = 1
    else
       FACTORIAL : = n*FACTORIAL(n-1)
end;
```

One potential problem involved in programming functions recursively is making sure that the execution of the algorithm eventually reaches the recursive exit; in the factorial example we require that the function is eventually called with a value of $n = 0$. If $n > 0$ in the initial invocation of the function, then repeated subtraction of 1 (see the "else clause") ensures this. But if this algorithm tried to execute with a negative value of n, say -5, the non-recursive exit would never be reached. Fortunately, if the invoking statement sends a negative value such as -5 to this factorial function, there will be a run time error because we have declared that the input parameter must be in the range from 0 to the maximum positive integer supported by that particular computer.

The run time error will cause an error message to be displayed. (The error message from the Pascal system may not be as straightforward as "You can't send a negative value to the factorial function." Such useful messages are obtained by having the function itself check on the appropriateness of independent variable values and generate messages specific to any problem encountered. But to do this, you would have to declare the input variable to be of integer type.)

Question 3.1.7

Try to get a square root of a negative number using your calculator or a computer program. What happens? Was the message specific to the square root function or a general system message?

An important issue concerning the recursive evaluation of n factorial is that there is also a non-recursive algorithm for evaluating n factorial:

```
function FACTORIAL(n: 0 .. maxint): 0 .. maxint;

var ans, i: 0 .. maxint;

begin
    ans : = 1;
    for i : = 1 to n do ans : = ans*i;
    FACTORIAL : = ans
 end;
```

This function uses an iterative or counting algorithm, not a recursive one, to obtain n factorial. It can be proved that it is always possible to find a non-recursive algorithm equivalent to a recursive one. Some of the considerations a programmer might use in deciding whether to use a recursive or non-recursive algorithm are discussed in **Chapter Five.**

Question 3.1.8

Can you write iterative evaluation algorithms for the recursively defined functions in this section?

Section 3.2
Recursive procedures

Now let us see how thinking recursively can be applied in situations other than evaluating functions. Many of the applications of recursion are found in procedures which perform tasks other than function evaluation.

Searching

Our first examples will be from the general category of searching for a particular item, called a search key, in a list of items stored in an array or list. For example, we look up a name (the search key) in a telephone book (the list) to get the phone number. In computer applications, an entire collection of associated data – for example, a person's name, address and telephone number – is often called a record and the part of the record which we check to see if it matches the search key is called the key field. To keep things simple here, we will think of the records as having two parts: a student's name and the student's grade point average. The part of the record containing the name will be the key field. We shall think of the records as arranged in a 1-dimensional array, one record per row. The purpose of the search is to tell us the grade point average of a given student whose name is supplied as the input search key. Of course, we will have to allow for the possibility that the search key (name) is not on the list.

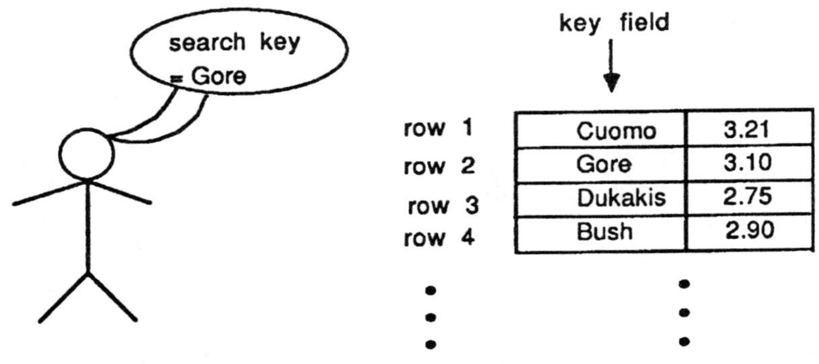

Figure 3.2.1

There are many ways to do searching. In fact, the study of different search algorithms is an important topic in computer science. One of the easiest searching techniques is called linear or sequential search. In this technique, the records in the array are examined one at a time in order to see if the key field matches the search key. If the entire array has been processed and no match is found, a message to this effect is printed out.

Our purpose here is not just to learn how to do linear search; our purpose is to see how linear search can be thought of as a recursive algorithm. We recall that our recursive functions had two parts, a general recursive one in which the problem at hand was solved in terms of a similar problem of smaller scope and non-recursive exit in which the answer was known. In looking for a recursive algorithm to carry out a process, we need to identify the same two parts: recursive and non-recursive. Generally speaking, this means that we have to try to split the original problem into two parts. One part is essentially the same as the original problem but smaller in scope. In the other part the answer must be apparent.

In doing a linear search, the first process is to compare the search key with the key field in row one. If the search key and the key field in row one do not match, what do we want to do? Search the rest of the list, of course. But how do we search the rest of the list? We start by comparing the search key to the key field in row 2. In doing this, are we not going through the same process we started with row one, with the important difference that the new problem is of smaller scope (since the number of rows to deal with is one less then before)? Aha! We have found the recursive part of the process. If a comparison does not result in a match then we initiate the same process over again with the unsearched, remaining portion of the data array.

The non-recursive exit part of the linear search must correspond to the circumstances under which we no longer have a reason to search. There are two such situations. When the search key is compared to the key field in the kth row, if there is a match, the answer to be returned is the grade point average in row k. Since the answer has been determined, there is no reason to continue; this is a non-recursive exit. The other situation which dictates that we stop searching is when there is no match anywhere on the list. How can this be detected? If the "remaining portion of the data array" is an empty list having no entries, then we have another non-recursive exit, the one which provokes the "not found" message.

We summarize this discussion of a recursive linear search with pseudocode for such a search. This code is designed to search the portion of an array between the row numbered "search position" and the row numbered "last position." Thus to search an entire array of 100 records, it should be sent the parameters 100 and 1 for last_position and search_position respectively. This pseudocode assumes that "datatype" and "arraytype" have been defined in the program which calls this procedure.

```
procedure LINEAR_SEARCH(last_position,
                 search_position: 1 .. maxint; search_key: datatype;
                 data_array: arraytype; var grade: real);


(* Searches  data_array from search_position to last_position,
looking for search_key and returns corresponding grade or  prints a
message if search_key is not on the list.  *)


begin
   if search_position <= last_position
     then
         (* there is an unsearched part of the array *)
         if search_key = key field of record in row
            whose number is search_position
           then
               grade := entry in grade field of record whose number is
                       search_position
         else
             LINEAR_SEARCH(last_position,
                         search_position+1,
                         search_key,
                         data_array,
                         grade)
     else
         writeln('search key not found in list')
end;
```

Question 3.2.1

Can you find the recursive part and the two non-recursive exits in this algorithm for a linear search?

Telephone chains

Here is an example of a process frequently carried out by people who have never heard of the word "recursion":

Each member of a volleyball team has the same list of team members and their phone numbers. The first person on the list gets a call telling the time and location of the next game. This person calls the second person on the list, who calls the third, and so on until all team members have the information.

Question 3.2.2

Can you write a recursive pseudocode algorithm to describe this process, which we can call PHONE_CHAIN_ONE?

There are, of course, other ways to carry out a phone chain. Let us specify a volleyball team of 15 people for this next method. Suppose that when the first person was given the job of informing the other 14 teammates about the next game that person called two people on the list and passes on half of the problem to each of them. Specifically, person 1 calls person 2 and passes on those people numbered 3 through 8 and also calls person 9 and passes on those people numbered 10 through 15.

As before, each of the teammates receiving the job handles it the same way as did the person passing on the job. That is, 2 calls 3 passing on 4 and 5, and calls 6 passing on 7 and 8; 9 calls 10 passing on 11 and 12, and calls 13 passing on 14 and 15. Of course, after 4, 5, 7, 8, 11, 12, 14, and 15 get their calls, the entire team knows the time and place of the next game; there is no list passed on to any of these players and they make no phone calls. No redundant phone calls were made and no player made more than two calls. And, as with the previous type of phone chain, we note that not only is this method clever, it is in widespread use by a lot of people who never heard of recursion.

A pseudocode algorithm for this second type of phone chain might look like this:

```
procedure PHONE_CHAIN_TWO( list: listtype);

begin
    if I am to pass the message on to anyone,
        i.e. my list is not empty
      then
        begin
            divide my list into two halves;
            call first person on first half;
            pass on message and first half of my list;
            call first person on second half;
            pass on message and second half of my list
        end
      else do nothing
end;
```

complex in this phone chain than in PHONE_CHAIN_ONE. In particular we see that the remaining job is being split or divided into two sub-jobs. Splitting the job into two roughly equal halves would be expected to speed up the time it takes to get the message spread. In PHONE_CHAIN_TWO it takes less time for everyone to get the message. Using the team of 15 people, we see that in PHONE_CHAIN_ONE, after the first person gets the message, 14 phone calls must be made before we reach a last person, who makes no calls. The first and last people are separated by 14 phone calls in PHONE_CHAIN_ONE. In PHONE_CHAIN_TWO, there are more "last" people, but each is separated from the first person by only 3 phone calls, as can be seen in **Figure 3.2.2**. This splitting technique is frequently used in computer science and has been named "divide and conquer."

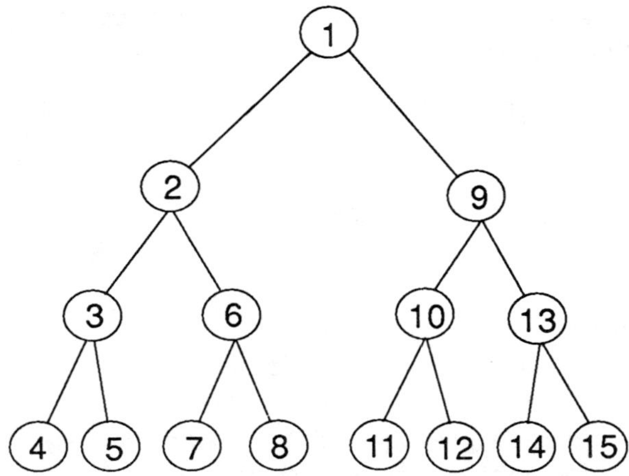

Figure 3.2.2 The pattern of calls in
phone_chain_two

Question 3.2.3

In an interactive computer program, such as one designed to give a young student drill in addition facts or one allowing a travel agent to interact with an airline's reservation data base, error-checking procedures are included to verify that the user's entry is of a valid form before that entry is processed. For example, 26 is a valid entry as an answer to an addition problem in arithmetic but 8t3 is not. Here is a recursive procedure, in pseudocode, which could be used for error checking. Can you think of any potential problems associated with using such a procedure?

```
procedure GET_ENTRY(var entry: entry_type);
(* entry will be returned if it is verified as valid *)


begin
    writeln ('please enter your response');
    readln (entry);
    if entry is not of valid form
        then
            GET_ENTRY(entry)
end;
```

Sorting: Insertion Sort

Sorting is a common activity in applications of computing. For example, suppose the Census Bureau wants to publish a list of major American cities. In particular, the cities may originally be in alphabetical order, or perhaps in random order, and the Census Bureau wants to put them in increasing order of population. Reordering a list so as to put the items in some particular order is called sorting. There are dozens of different algorithms to sort data. These algorithms differ in how fast they are if the data is randomly arranged, in how much storage they require, and in how efficient they are if the data is almost in order already. Sorting is a major topic in data structures texts (see, for example, Tenenbaum and Augenstein [1986].) For simplicity, we will consider only the key field of the record.

One kind of sort procedure, called the insertion sort, is similar to the way many people arrange their cards in a card game. As the cards are being dealt, a person picks up one card at a time, scans the cards in the hand already, and inserts the new card into the proper position. In arranging a hand of cards using an insertion technique, we rely on the hand we are holding as being already arranged. In fact we know that it was arranged by previous uses of the insertion method: Recursion is at work.

We shall develop a recursive computer algorithm to mimic this familiar card-sorting process. For simplicity, assume that we are dealing with integers which start out in an array in some unspecified order. When we are done sorting, they must be in numerical order in the array, starting from the smallest and ending with the largest. Whereas in sorting cards we have two different locations for the cards, the table and the hand, in our computer sort we will always have all the integers already stored in the array. At any given stage of the process there will be a front part of the array where the items are in sorted order and a back part of the array where integers are still in their original order. The front part of the array corresponds to the hand and the back part corresponds to the table. Given this condition of the array, partially sorted:

sorted part of array   unsorted part of array
3, 5, 8, 11, 20, 21, 42, **|** 23, 1, 9, 31, 27, 6, 39, 14

after one insertion we have:

sorted part of array   unsorted part of array
3, 5, 8, 11, 20, 21, 23, 42, **|** 1, 9, 31, 27, 6, 39, 14

A recursive algorithm requires that any one stage of the process be solvable in essentially the same way as the previous stages and that somewhere there is a non-recursive exit. Here's the thinking that leads us to a recursive algorithm for insertion sort. The whole array would wind up sorted if we could carry out two steps:

1. sort the first $n - 1$ items
2. insert the remaining item (the one in position n) into its proper place among the already sorted items.

Step 1 is done by a recursive call to the insertion sort algorithm provided $n > 1$. If $n = 1$ then the data list has just 1 item and is already in order, so nothing at all is done. This is the non-recursive exit.

In order to write an algorithm for this insertion sort, it is convenient, and in keeping with a top-down approach, for us to assume the existence of a procedure which does the actual inserting of a chosen item into an array of already sorted items. Of course, we need to know what parameters this procedure needs, exactly what results we can expect it to produce, and its name. We will call the procedure INSERT. This procedure will have as its parameters the array, which we will call data_array, and a positive integer k. INSERT will assume that the items from 1 to $k - 1$ of data_array are already sorted. It will traverse the items of data_array to determine where the kth item is to be inserted, and then do the actual insertion, thus creating a new sorted version of the first k items of data_array. The details of how this is done are not of importance to us now as long as we understand that after INSERT has done its job, the first k items of data_array are in the proper order.

We noted that the final stage of insertion sort is the insertion of the nth item into the data_array of the first n - 1 items, which already are sorted. The words "which already" tell us that the call to INSERT must come after the recursive call to the insertion sort itself.

The parameters to the insertion sort must include data_ array, the array of data records to be sorted, and n, the number of records to be sorted, the understanding being that they are the first n elements of data_array. In many cases we want to sort the entire array, so n is the index of the last item in the array. However, we might want only to sort the beginning part of an array, so n does not need to be the size of the array. Pascal pseudocode for this algorithm follows:

```
procedure INSERT_SORT(n: 1 .. maxint; var data_array: arraytype);

var  k: 1 .. maxint;

procedure INSERT(k: 1 .. maxint; var data_array: array_type);

    (* details of INSERT omitted *)

begin  (* INSERT_SORT *)
    if n > 1
      then
        begin
            k : = n;
            INSERT_SORT(k-1, data_array);
            INSERT(k, data_array)
        end
end;  (* INSERT_SORT *)
```

Let us trace this routine to see how it does its work. For our example n will be 8 and data_array will start with the eight integers 14, 3, 27, 52, 46, 18, 21, 22, in that order. The last item, item n, is positioned after the first n - 1 items are sorted; to get those n - 1 items sorted (the recursive call) the computer must position item n - 1. But that presupposes that the first n - 2 items are in order, and we back our way up the array until, in fact, the computer starts the actual work by positioning the first item. Although the recursive calls occur for decreasing values of n as execution progresses, the actual insertions proceed in the opposite order. The table below shows the insertion of the items in place in data-array starting with the positioning of the first item. A vertical line is used to separate the portion of data-array already sorted (the beginning of the array) from the portion containing items left to be inserted into place.

index of item being positioned	contents of data array
1	14,	3, 27, 52, 46, 18, 21, 22
2	3, 14,	27, 52, 46, 18, 21, 22
3	3, 14, 27,	52, 46, 18, 21, 22
4	3, 14, 27, 52,	46, 18, 21, 22
5	3, 14, 27, 46, 52,	18, 21, 22
6	3, 14, 18, 27, 46, 52,	21, 22
7	3, 14, 18, 21, 27, 46, 52,	22
8	3, 14, 18, 21, 22, 27, 46, 52	

We now describe the algorithm for the INSERT procedure. The basic idea is to locate where to position the new item by comparing it with the sorted items, taken in reverse order, and shifting down, or to the right, previously sorted items which are bigger than the new item in order to make room for the new item. For example, consider the problem of doing the last insertion involved in the example described above, i.e. the re-positioning of the number 22 in the array:

$$3, 14, 18, 21, 27, 46, 52, \; | \; 22$$

First we store the 22 in a temporary location, in effect creating an unused space at the end of the array. We successively compare the 22 stored in the temporary location to the numbers 52, 46, 27, etc. When we do the comparison with 52, since 52 is larger than 22, we move the 52 rightward one location to where the 22 was. Next we compare 22 with 46; since 46 is larger, we move the 46 to the right. We continue to compare and move numbers to the right until the number in the sorted portion of the array, here the 21, is not greater than the number in the temporary location, in our case the 22. At this point we view the situation like this:

temporary location: 22
array: 3, 14, 18, 21,   , 27, 46, 52

The 22 is then placed immediately following the 21 (the original position of the 27). This gives us the array in the proper order.

Question 3.2.4

Can you write pseudocode or Pascal code for INSERT? Can you make this code recursive?

Question 3.2.5

Can you see how to modify the procedure INSERT and the recursive procedure INSERT_SORT so they sort only the data items having indexes from j to n, not the whole array data_array?

A pattern for recursive algorithms

The recursive algorithm insertion sort fits into a general pattern for recursive algorithms:

```
initialize as needed;
   if this is not the non-recursive exit
   then
      do what can be done now and make recursive calls as
      needed
   else
      carry out the operations associated with the non-
      recursive case;
```

Other examples of this pattern were seen earlier in this section. The linear search was a special case, with an extra non-recursive exit on the **then** branch.

Question 3.2.5

What are the initialization, recursive, and non-recursive parts of each of the recursive procedures we have seen so far?

Question 3.2.6

There is a common search method that requires that the entries in the array be ordered. When the array is sorted, then a "binary" or "half-interval" search proceeds by first seeing if the middle item in the array matches the search key. If so, the search is over. If not, then because the array is sorted, we can tell in which of the two halves of the array to continue the search. Of course, the search in one of the two halves is done in exactly the same way as the original search; that is, recursion is at work here. Can you write a pseudocode algorithm for a recursive binary search?

Question 3.2.7

In **Section 2.2** there was a syntax diagram for algebraic expressions. Can you write pseudocode for an algorithm to determine whether a string of characters is a legal algebraic expression – can be built up using the syntax diagram? *Hint*: the easiest way to write the algorithm is to use recursion.

Chapter Four
Applications of Recursion

In this chapter we look at some procedures for which it is quite easy to develop algorithms which are recursive and not at all obvious how to construct equivalent non-recursive algorithms. Our applications are merge sort, symbolic differentiation, and an artificial intelligence technique called backward chaining.

Section 4.1
Sorting: Merge Sort

In many computer applications, two sorted arrays are to be merged into a third sorted array. Simply by comparing the "next" items in each of the two short lists, it is easy to determine which next item – namely the smaller – should be chosen as the next item in the longer list being formed. For example, suppose we are trying to create Merged Array in **Table 4.1.1** out of Array One and Array Two. We begin by comparing the 5 and the 3, the first items in the shorter arrays, and deciding which to place into Merged Array. After placing the three, the "next" items in the two arrays are 5 and 4. These are compared and the smaller is placed in the first empty space of Merged Array. The process continues till either Array One or Array Two is out of elements. At this point all of the remaining elements in the non-empty array are placed into Merged Array. Most of the work in the merging is keeping track of "next" items. We will call this algorithm MERGE (not to be confused with MERGE_SORT which is based on it). Pascal code for MERGE is given at the end of this section. Our interest is in applying the idea of merging to the problem of sorting an array.

Array One	Array Two	Merged Array
5	3	3
6	4	4
7	9	5
13	11	6
		7
		9
		11
		13

Table 4.1.1

In MERGE_SORT the data items which need to be sorted are all in one original array. The idea of the algorithm is to break this array up into two approximately equal sub-arrays. For example, if the original array has eight entries, then we would name as "first half" the subarray with subscripts 1 through 4 and we would name as "second half" the subarray with subscripts 5 through 8. Now if only these two sub-arrays were sorted, we could think of them as Array One and Array Two in the previous example and use MERGE to sort them into a new array. Then this array could be copied back into the original. The only thing which stops us from doing this is that the sub-arrays are not sorted. But we can get them sorted by making a recursive call to the very algorithm we are developing, MERGE_SORT. A picture of this situation is shown in **Figure 4.1.1**:

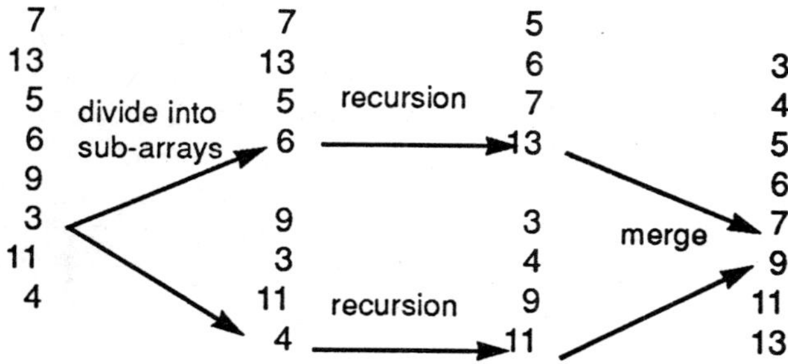

Figure 4.1.1 How merge-sort works

The non-recursive exit occurs when there is an array of fewer than two items, since in that case there is no sorting to be done.

We give the code for MERGE_SORT below. Notice that it depends on MERGE whose declaration within MERGE_SORT we have abbreviated since it is rather long (see the end of this section). The MERGE procedure will need four parameters: data_array, the array containing the unsorted data items; low, the subscript of the first item of the "first half" subarray; mid, the subscript of the last item in the "first half"; and high, the subscript of the last item in the "second half." Since we are arranging things so that the second half subarray immediately follows the first half subarray within the array, we do not need a parameter for the subscript of first item of the second half; it is mid + 1.

```
procedure MERGE_SORT(low, high : 1 .. maxint;
                 var data_array : arraytype);

var mid: 1 .. maxint;

procedure MERGE;   (* see end of section *)

 begin    (* MERGE_SORT *)
     if low < high   (* list has more than one item *)
        then
          begin
              (* find where to split the array *)
              mid := div(low+high,2);
              (* send first half of list to be sorted *)
              MERGE_SORT(low, mid, data_array);
              (* send second half of list to be sorted *)
              MERGE_SORT(mid+1, high, data_array);
              (* merge the two sorted halves *)
              MERGE(low,mid,high, data_array)
          end
 end;   (* MERGE_SORT *)
```

It may not be obvious that the execution of this recursive procedure will always reach a non-recursive exit, but it is true. Each time recursive calls are made to merge_sort, the values of either high or low have been changed to make the length of the new array to be sorted approximately one half of the length of the array in the previous call of merge_sort. On each round of recursive calls, the low and high indices of the array list to be sorted on that call are closer to each other than they were on the previous round. Eventually we reach the condition that low ≥ high, which is the non-recursive exit. There is no work to do if the list contains fewer than two items.

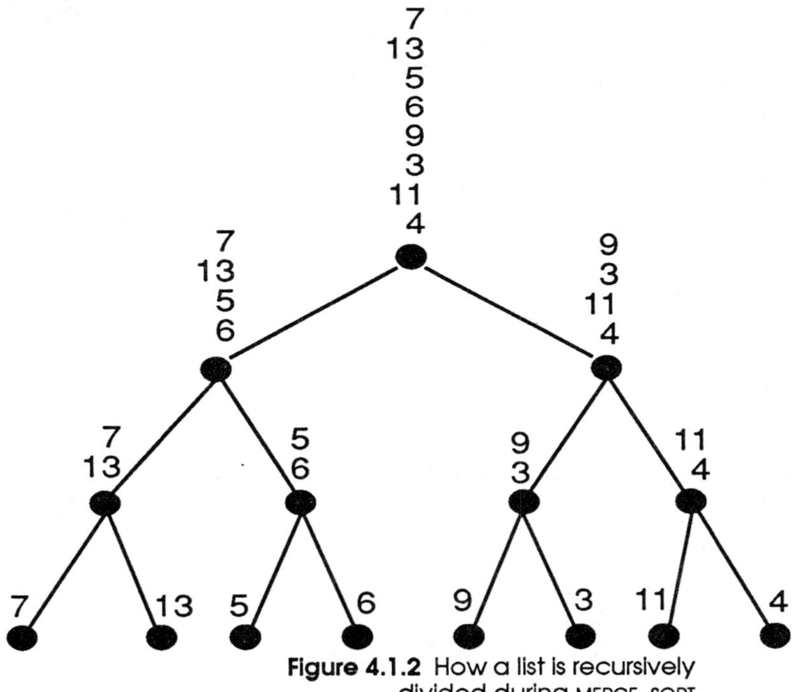

Figure 4.1.2 How a list is recursively divided during MERGE_SORT

Figure 4.1.2 shows how the original list is successively divided during the various calls to merge_sort. The top node of the tree represents the data sent to merge_sort initially. merge_sort divides the list and then makes two recursive calls. These recursive calls, and the sublists they receive are represented by the two nodes below the top node. The recursive calls these two versions of merge_sort make are represented by the four nodes on the next level, etc. No merging actually gets done till we get to the bottom level of the tree. The sequence of merges can be followed in **Figure 4.1.3** by going up the tree, merging smaller lists into larger ones. For example, first the one-item lists consisting of 7 and 13 are merged into one ordered list consisting of 7 followed by 13.

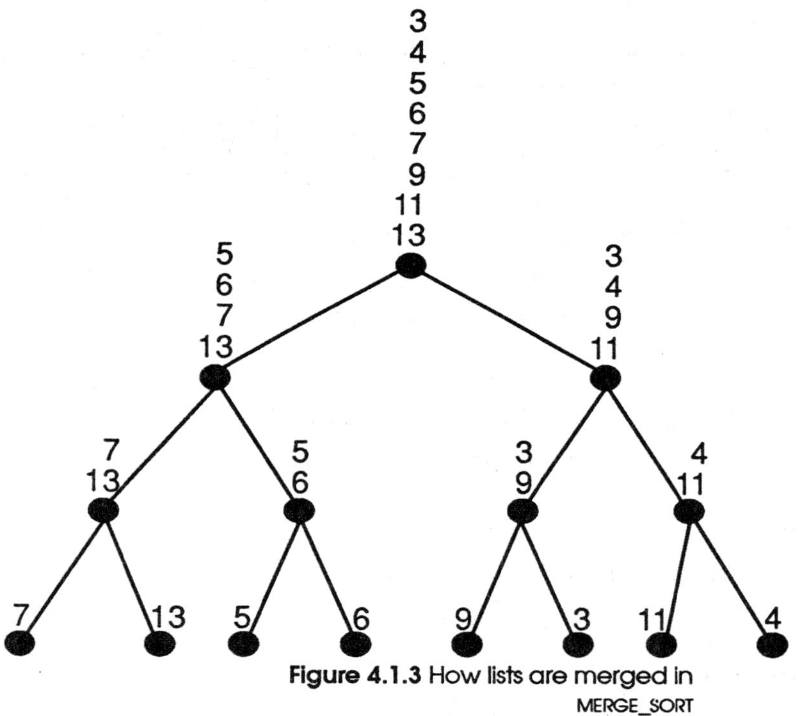

Figure 4.1.3 How lists are merged in
MERGE_SORT

Let us stop here a moment to look at the power of recursion. In the case of MERGE_SORT, for example, we have created in short order an algorithm for a relatively complex process. Even granting that we have postponed the details of the MERGE sub-algorithm, what we have done seems almost magic. By assuming that MERGE_SORT works on a short array we can get it to work on a larger one, with the help of MERGE. Ironically, it is perhaps easier to understand this top-down approach to MERGE_SORT than to actually trace what happens when it processes a specific set of data. Because the algorithm we are using in MERGE_SORT repeatedly halves the size of the task at hand, the length of the array to be sorted, it is a divide and conquer algorithm.

Question 4.1.1

Can you name any algorithms presented in earlier chapters that are divide and conquer algorithms?

Before we leave our discussion of MERGE_SORT, we have one job left to do. Here is code for the MERGE procedure. It merges the two short contiguous subarrays in the array data_array into one long list in the array temp and then replaces the contents of the short subarrays by their merger from temp. The parameters low, mid, and high indicate the positions of the two subarrays, as described earlier.

```
procedure MERGE(low, mid, high: 1 .. maxint;
                    var data_array: arraytype);


var   index1, index2,
      temp_index, array_index:   1 .. maxint;
      temp:            arraytype;


begin
  (* initialize counters *)
  index1 : = low;
  index2 : = mid + 1;
  temp_index : = 1;
  (* merge  until one subarray is empty*)
  while (index1 <= mid) and (index2 <= high) do
  begin
      if data_array[index1] < data_array[index2]
        then
          begin
            (* draw from first short array *)
            temp[temp_index] : = data_array[index1];
            index1 : = index1 +1;
          end
```

```
          else
            begin
              (* draw from second short array *)
                temp[temp_index] : = data_array[index2];
                index2 : = index2 + 1;
              end;
          temp_index : = temp_index + 1
      end;   (* main while loop *)


(* one subarray is exhausted, copy remaining elements
 from other subarray into temp — only one of the
 following two while loops will execute *)

while index1 <= mid do
    begin
      (* copy from first half *)
        temp[temp_index] : =  data_array[index1];
        temp_index : = temp_index + 1;
        index1 : = index1 +1
    end;
  while index2 <= high do
    begin
      (* copy from second half *)
        temp[temp_index] : =  data_array[index2];
        temp_index : = temp_index + 1;
        index2 : = index2 +1
    end;
  (* move items from temp back to data_array *)
  temp_index : = 1;
  for array_index : = low to high do
    begin
        data_array[array_index] : = temp[temp_index];
        temp_index : = temp_index +1
    end
end;     (* MERGE *)
```

Both the merge sort and the insertion sort proceed by splitting the unordered array into two parts in a mechanical and thus easy fashion. In both sorts, the hard work is putting the two halves back together. (For example, compare how much code MERGE contains compared to the amount of code in the main routine MERGE_SORT.) One way of classifying sort routines is by how easy or hard the splitting and joining processes are. In this type of classification, both insertion and merge sorts are "easy-split-hard-join" (Merritt, 1985).

Question 4.1.2

Another sorting procedure which is usually described recursively is called quick sort. It might be called a hard-split-easy-join type of sort. The splitting involves rearranging the array elements so that somewhere, hopefully near the middle of the array, there will be a "partitioning element" with the following property: All elements in the array prior to the partitioning element are less than or equal to the partitioning element and all elements in the array after the special element are greater than or equal to the partitioning element. Achieving this much can be regarded as a rough kind of sort. Each of the two halves is then refined by another application of the quick sort. Much has been written about quick sort; one of its most interesting aspects is the choosing of the partitioning element. Can you develop an algorithm for a quick sort? *Hint*: begin by devising a way to take the first array element, data_array(1), and performing appropriate comparisons and interchanges involving various array members till data_array(1) winds up being a partitioning element. The challenge is to write efficient code to reposition data_array(1) properly so that all elements before it in its new placement are less than or equal to it and all elements after it are greater than or equal to it. Tenenbaum and Augenstein (1986) has a nice discussion of quick sort, as do most other books on data structures or on algorithms.

Section 4.2
Taking Derivatives by Recursion

In a calculus course, one of the first topics is taking derivatives of functions. A derivative of a function is another function whose properties can tell us much about the nature of the original function. For example, instead of graphing the function, we can often use the derivative to tell at which values of x the original function is increasing and at which it is decreasing. Working from the theory, one develops rules for how to take derivatives, so that after one learns the rules, the process of taking derivatives becomes more or less mechanical. In this section, we review those rules and note that applying them involves a form of recursion. We develop a recursive algorithm for taking derivatives of a restricted class of functions. Since this algorithm will not "understand" differentiation or "see" the function in any way, but rather just process a string of characters which express the function, we say that this is an algorithm for "symbolic differentiation." In this phrase, the word "symbolic" connotes that the computer is finding a derivative in a purely mechanical way, without "understanding" calculus. The word "symbolic" does not connote that the derivative that is found is in any way different from the one that would be obtained by a human being.

Defining the functions

We will be restricting our attention to one category of functions: those which can be written as expressions described by the syntax diagram below.

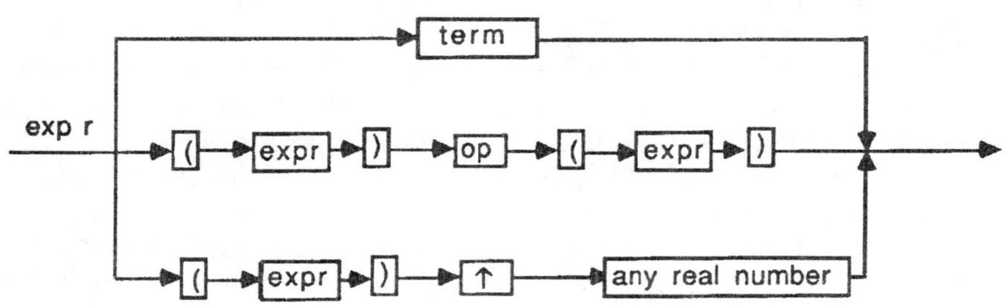

We use "term" to stand for the variable x or for any real number, such as 714, possibly with a decimal point and possibly preceded with a + or - sign. We use "op" to stand for any of the symbols +, -, *, /, which express the four basic operations of arithmetic. The "up-arrow" denotes exponentiation. For example, x^2 would be written $x \uparrow 2$, thereby keeping all the symbols on the same line.

Question 4.2.1

Can you develop a syntax diagram for a term?

The syntax diagram for an expression allows us to build expressions like:

```
3.1
x
(3.1) + (x)
( (3.1) + (x) )↑4
```

Notice that this syntax diagram produces expressions with lots of parentheses, more than we would normally use in our own pencil and paper work. The expression $(3.1 + x)\uparrow 4$ cannot be produced by this syntax diagram although the equivalent expression $((3.1) + (x)) \uparrow 4$ can. Requiring this excessive number of parentheses makes it easier for us later when we sketch code for doing symbolic differentiation.

Rules for taking derivatives

Below we present those rules for differentiation that we will use in this section. If you have not taken calculus or have not yet learned all these rules, you can still follow our discussion by adopting the same sort of symbolic approach that our algorithm will take: Given the form of a function, then the rules give the form of another function, which happens to be called a derivative. For our purposes, it is not necessary that you understand where the rules come from or the usefulness of the derivative, although such knowledge is of course very important. In the rules we use the abbreviation "expr" to stand for an expression in the formula for a function, and we use

$$\frac{d(\text{expr})}{dx}$$

as the notation signifying the derivative of "expr."

Basic Rules: $\dfrac{d(\underline{\text{any number}})}{dx} = 0$

$$\frac{d(x)}{dx} = 1$$

Sum Rule: $\dfrac{d(\text{expr1 + expr2})}{dx} = \dfrac{d(\text{expr1})}{dx} + \dfrac{d(\text{expr2})}{dx}$

Examples:

$$\frac{d(2 + x)}{dx} = \frac{d(2)}{dx} + \frac{d(x)}{dx} = 0 + 1 = 1$$

$$\frac{d(2x)}{dx} = \frac{d(x + x)}{dx} = \frac{d(x)}{dx} + \frac{d(x)}{dx} = 1 + 1 = 2$$

Question 4.2.2

Can you use the second example above to show that a is the general form for the derivative of $a*x$, where a is a positive integer?

Difference Rule: $\dfrac{d(\text{expr1 - expr2})}{dx} = \dfrac{d(\text{expr1})}{dx} - \dfrac{d(\text{expr2})}{dx}$

Example:

$$\frac{d(0 - 6x)}{dx} = \frac{d(0)}{dx} - \frac{d(6x)}{dx} = 0 - 6 = -6$$

Question 4.2.3

Can you use the result of the **Question 4.2.2** and the example above to show that the derivative of $a*x$ is a, when a is any integer including negative integers?

Product Rule:

$$\frac{d(\text{expr1} * \text{expr2})}{dx} = \text{expr2}*\frac{d(\text{expr1})}{dx} + \text{expr1}*\frac{d(\text{expr2})}{dx}$$

Examples:

$$\frac{d((x\uparrow 2)}{dx} = d(x * x) = x * 1 + x * 1 = 2x$$

$$\frac{d(3.2\ x)}{dx} = \frac{d(3.2 * x)}{dx} = x * 0 + 3.2 * 1 = 3.2$$

Question 4.2.4

Can you use the first example above to convince yourself that a general form for the derivative of $x\uparrow n$ is $n*x\uparrow(n\text{-}1)$, where n is any positive integer? *Hint*: First try $n = 3$.

Question 4.2.5

Can you use the second example above to show that in general, the derivative of $a*x$ is a, where a is any number, not necessarily an integer?

Question 4.2.6

Can you use the results of the two previous questions to verify the rule below, where a is any number and n is a positive integer:

$$\frac{d(a*x\uparrow n)}{dx} \qquad = \qquad n*a*x\uparrow(n - 1)$$

Quotient Rule:

$$\frac{d(expr1\ /expr2)}{dx} = \frac{expr2*\frac{d(expr1)}{dx} - expr1*\frac{d(expr2)}{dx}}{(expr2)\uparrow 2}$$

Example:

$$\frac{d((2x)/(3x\uparrow 6))}{dx} = \frac{(3x\uparrow 6)*(2) - (2x)*(18x\uparrow 5)}{(3x\uparrow 6)\uparrow 2}$$

$$\frac{d(x\uparrow(-1))}{dx} = \frac{d(1/(x\uparrow 1))}{dx} = \frac{(x\uparrow 1)*(0) - 1*1}{(x\uparrow 1)\uparrow 2} = \frac{-1}{x\uparrow 2}$$

$$= \frac{-1*x\uparrow(-2)}{x\uparrow 2}$$

(Notice that in this example we have occasionally left

*out asterisks when it is clear that multiplication is intended - i.e., we write 2x instead of 2*x. Later in this section we will outline computer programs to do differentiation, and these programs will not work on expressions where asterisks have been left out on the assumption that the reader knows where they belong.)*

Question 4.2.7

Can you use the result of **Question 4.2.6** and the second example above to expand the rule given in **Question 4.2.6** to cover all integer values of n? (In fact, the rule holds if the exponent is any real number, but showing that is beyond the scope of this section.)

Power Rule:

$$\frac{d(\text{expr}\uparrow p)}{dx} = p*\text{expr}\uparrow(p - 1)* \frac{d(\text{expr})}{dx}$$

where p is any real number.

Example:

$$\frac{d((2x\uparrow 5 - 4.1x\uparrow(-3))\uparrow 7)}{dx} =$$

$$7(2x\uparrow 5 - 4.1x\uparrow(-3))\uparrow 6* (10x\uparrow 4 + 12.3x(-4))$$

Question 4.2.8

Show that the form of the derivative of $a*x\uparrow n$ can be found using the power rule.

As you can see, except for the basic rules, the rules for taking derivatives of expressions are stated in terms of taking derivatives of other related expressions. This means that the process of taking derivatives is recursive. For example, in finding the derivative of a sum of two monomials, we needed to find the derivative of each monomial. Each of those derivatives, in turn, may fall under the power rule special case or under one of the two basic rules. In general, derivatives of complicated expressions are formed from some combination of the derivatives of their component expressions.

Question 4.2.9

Can you use the rules to find the derivatives of these functions? $4x{\uparrow}5 + 17x{\uparrow}(-2)$, $3 - x{\uparrow}(3)$, $(4x{\uparrow}2)*(-9x{\uparrow}8)$, $(7x{\uparrow}6)/(2x{\uparrow}4)$, $(3x{\uparrow}2){\uparrow}8$. The last three functions can be simplified algebraically first, and the derivative taken of the simplified function. Of course, the derivatives in both cases should be algebraically equal.

Question 4.2.10

Here are some more complicated functions to use for practice in taking derivatives: $(7x{\uparrow}2 + 23x{\uparrow}(-5))*(4x{\uparrow}7 - 2x{\uparrow}9)$, $(6x{\uparrow}5 + 8x{\uparrow}4){\uparrow}27$, $(23x{\uparrow}40 - x{\uparrow}8)/(275 + x{\uparrow}5)$. In each of these, the expressions in the formulas for the derivative are sufficiently complex that more than one rule must be invoked before arriving at expressions for which a basic rule is applicable.

Do not lose heart if you are uncertain about following the rules. After studying the development of the algorithm to do symbolic differentiation, the entire process should become much clearer.

Computer programs to do differentiation

Recently, computer software has been written to take derivatives of algebraic expressions such as

$$(3.1 + x) \neq 4$$

These programs are often called symbolic differentiation programs. To use such a program, one types in the expression one character at a time into a variable which is an array of characters — a so-called string variable. In the case of the above expression, the characters are first the "(", then "3", etc., followed by some special character to indicate that the end of the expression has been reached. The program then does some manipulations involving the string and outputs a string of characters that represents the derivative. For this example the answer string would be

$$4*(3.1 + x)\uparrow3$$

Did you notice that the power rule would be applicable here and that the derivative of 3.1 + x is 1, so it "does not appear"?

Symbolic differentiation programs such as those contained in MACSYMA, MAPLE and MATHEMATICA can differentiate expressions in a blink of an eye that would be difficult and time-consuming for people to get right. What has been taught for hundreds of years as a human skill is now often better done by machine. Of course it's still important for scientists and mathematicians to know what a derivative means and what the applications of derivatives are.

Although these symbolic differentiation programs are a major advance, they are based on some fairly simple rules, some of which we have listed above, and the use of recursion. In the remainder of this section, we shall present a description of how our small category of expressions can be differentiated with the help of recursive procedures.

Parse Trees

When we build an expression using our syntax diagram **(Figure 4.2.1)**, we can conveniently record the history of how we built the expression with a diagram called a parse tree or an expression tree. For example let's build ((3.1) + (x))$\uparrow$4 using the syntax diagram and let's build the tree at the same time.

Step 1. create the expressions   create nodes of a tree
        3.1 and x representing 3.1 and x

Step 2. combine (3.1) and (x) with the op + , using the middle branch of the diagram

add a node labeled with + and connect it to the nodes for 3.1 and x

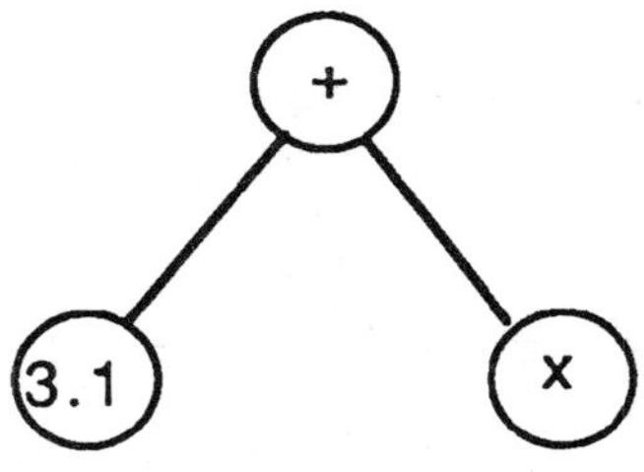

Step 3. create the expression ((3.1) + (x))↑4 using the middle branch of the

create a node for the integer 4 and a node for ↑ and connect ↑ to the syntax diagram node for 4 and to the + node representing ((3.1) + (x))

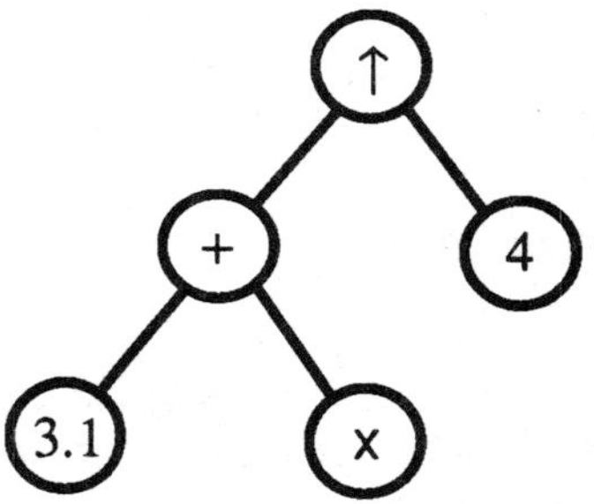

Although an expression tree may seem an odd way to describe an algebraic expression, it is actually very natural. The order of operations is quite clear and there is no need for parentheses. It is probably only the desire to have algebra appear like ordinary writing, i.e., a left to right string, that accounts for the fact that we don't use expression trees in our everyday use of algebra. But in our algorithmic analysis of algebraic expressions, it is best to be released from this tyranny of typography.

Outline for the algorithm to build a parse tree

Building a parse tree to represent an expression is called *parsing* the expression. Although we have built trees by hand, it can also be done automatically by software. (You can learn more about parse trees in a data structures text such as Kruse or Tenenbaum and Augenstein.) In other words one can write software that accepts as input the string of symbols representing an expression – the thirteen symbols making up ((3.1) + (x))↑4 in our example – and calculates a data structure (a representation in the computer) for the tree shown above. We won't deal any further with parsing. We will assume that we have a procedure that takes an expression as input and returns a tree as output, i.e.:

procedure PARSE(input_exp: string, **var** parse_tree: tree);

We will show how to carry out operations on the parse tree that result in a tree that represents the derivative of the original expression. These operations will be contained in the following procedure:

procedure DERIVATIVE_OF_TREE (parse_tree: tree; **var** deriv_tree: tree);

We will also need to assume that we have a procedure available that reverses this parsing process, i.e., it takes a tree representing an expression and it outputs the expression in string form, i.e.:

procedure UNPARSE(deriv_tree: tree; **var** deriv_exp: string);

We combine these procedures into a procedure for differentiating an expression represented by a string as follows:

procedure DERIVATIVE_OF_STRING(input_exp: string; **var** deriv_exp: string);
(* *declare parse_tree, deriv_tree and 3 sub-procedures* *)

```
begin
     PARSE(input_exp, parse_tree);
     DERIVATIVE_OF_TREE(parse_tree, deriv_tree);
     UNPARSE(deriv_tree, deriv_exp)
end;
```

Non-recursive exit for the algorithm

The algorithm DERIVATIVE_OF_TREE for differentiating an expression, given its parse tree, operates by recursion. We will start by describing the non-recursive exit, i.e., the cases where the derivative can be written down immediately without a recursive call.

When we study our rules for derivatives, we see that only the two basic rules give specific answers, as opposed to answers in terms of other derivatives. These rules must correspond to the non-recursive exit we desire. Thus we have: If a tree consists of a single node containing a number, the derivative tree is the tree consisting of a single node containing zero. If a tree consists of just one node containing an 'x', the derivative tree is the tree consisting of a single node containing a '1'. For brevity of expression we are not carefully distinguishing here between a number, such as one, and the character string representing that number, which is really what is stored in the node of the tree.

Now that we have found our non-recursive exit, we can begin writing our algorithm:

```
procedure DERIVATIVE_OF_TREE (parse_tree: tree; var deriv_tree: tree);

begin
   if parse_tree has just one node
```

```
then
    case node_contents of
        any number :  deriv_tree has 1 node with '0';
        'x'            :  deriv_tree has 1 node with '1'
    end
else
    (* This is where the recursion will be done - read on! *)

    -

    -

    -

end;
```

To carry out what we have described so far, we would have to be able to tell when a string corresponding to a node represents a number. This can be done by checking its characters one by one and seeing that all of them are digits except for possibly one which is a decimal point, and except for possibly a minus sign at the start of the string.

Some details of the algorithm

Now suppose the parse tree is not one of the easy ones for which the non-recursive exit handles the problem. Then it follows that the tree can be decomposed into a left subtree and a right subtree and an operation (op) connecting them. Consider, for example, the expression $(3.1) + (x)$. In this case, the general form is:

$$\text{input\_expr} = (\text{left\_addend}) + (\text{right\_addend})$$

which we describe in tree form as

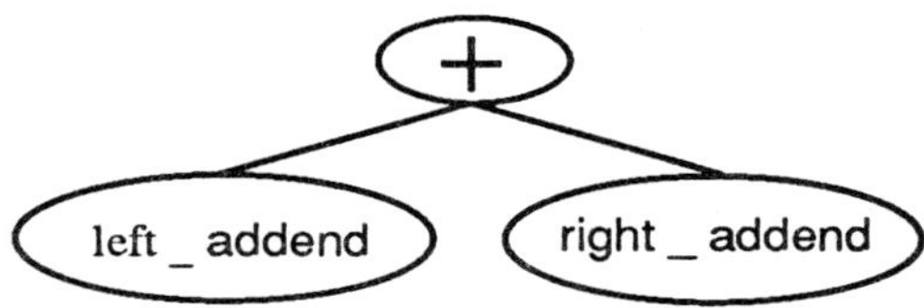

We recall the differentiation rule for sums of expressions:

$$\frac{d((\text{expr1}) + (\text{expr2}))}{dx} = \frac{d(\text{expr1})}{dx} + \frac{d(\text{expr2})}{dx}$$

This means, in tree terms, that to differentiate the previous tree requires us to build the following tree:

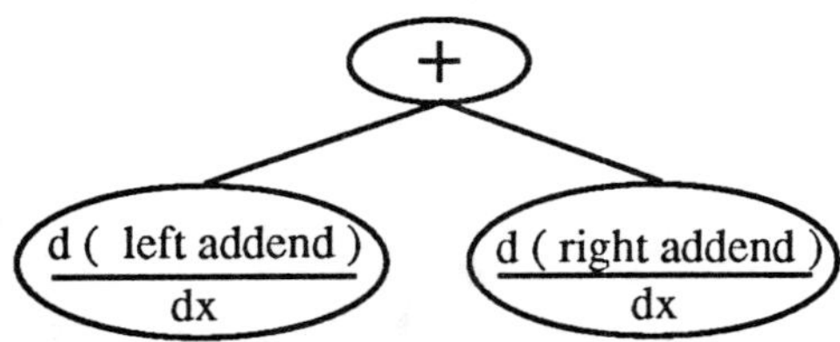

This is recursive in that it expresses the derivative in terms of the derivatives of two simpler expressions.

To carry out the appropriate tree operations so as to differentiate a tree, it is convenient to have the following tree-handling utilities:

procedure CHOP_LEFT(input_tree: tree; **var** left_subtree: tree);
(* *this procedure outputs the left subtree of the*
   top node of the input tree *)

procedure CHOP_RIGHT(input_tree: tree; **var** right_subtree: tree);
(* *this procedure outputs the right subtree of the*
   top node of the input tree *)

procedure CONNECT(left_tree, right_tree: tree;
                   op: **char**; **var** bigger_tree: tree);
(* *this joins two trees by hanging them off a node labeled op* *)

With the help of these utilities, we can express procedure SUM as follows:

procedure SUM (input_tree: tree, **var** deriv_tree: tree);
  (* *declare various trees and sub-procedures* *)

begin

    CHOP_LEFT(input_tree, left_tree);
    CHOP_RIGHT(input_tree, right_tree);
    DERIVATIVE_OF_TREE(left_tree, deriv_left_tree);
    DERIVATIVE_OF_TREE(right_tree, deriv_right_tree);
    CONNECT(deriv_left_tree, deriv_right_tree, '+',

                        deriv_tree)

end;

       Now suppose the op at the top of the parse tree is not +, but one of the remaining four possibilities: -, *, /, and ↑. For each of these cases, the appropriate rule for taking derivatives gives rise to a procedure that has to be included in the algorithm DERIVATIVE_OF_TREE. We shall call these procedures DIFFERENCE, PRODUCT, QUOTIENT, and EXPONENT. We now present the final structure of DERIVATIVE_OF_TREE with some functions and conditions, such as top_of_tree, left as pseudocode:

procedure DERIVATIVE_OF_TREE (parse_tree: tree; **var** deriv_tree: tree);
  (* *declare various sub-procedures* *)

begin

    if parse_tree has just one node
     then
       case node_contents **of**
         any number   : deriv_tree has 1 node with '0';
            'x'   : deriv_tree has 1 node with '1'
      end
     else
       case top_of_tree **of**
         '+' : SUM(parse_tree, deriv_tree);
         '-' : DIFFERENCE(parse_tree, deriv_tree);
         '*' : PRODUCT(parse_tree, deriv_tree);
         '/' : QUOTIENT(parse_tree, deriv_tree);
         '↑' : EXPONENT(parse_tree, deriv_tree)
       end
end; (* *DERIVATIVE_OF_TREE* *)

More details of the algorithm

The algorithm for EXPONENT is slightly more involved than the algorithm for sum. In thinking about EXPONENT, it is helpful to keep the example of the expression $((3.1) + (x)) \uparrow 4$ in mind. The tree for this expression has a top node containing $\uparrow$, a left subtree representing $(3.1) + (x)$ and a right subtree representing 4 and a node labeled $\uparrow$ connecting them. This decomposition into subtrees corresponds to thinking of $((3.1) + (x)) \uparrow 4$ as having the form

$$(\text{expr}) \uparrow 4$$

The power rule of differentiation says that

$$\frac{d}{dx}((\text{expr}) \uparrow n) = n * \text{expr} \uparrow (n\text{-}1) * \frac{d(\text{expr})}{dx}$$

Notice that again we have a rule that gives us the derivative in terms of another derivative. This recursion in the differentiation rule translates perfectly into recursion in our code for procedure EXPONENT, which will have a call to DERIVATIVE_OF_TREE.

To understand the algorithm we will outline for EXPONENT, it is helpful to recall the mechanical steps we human beings go through in order to differentiate $\text{expr} \uparrow n$. We first copy down the exponent n. We follow it by a multiplication sign $*$ and next to this we copy the expression expr and we raise it to the power obtained by subtracting 1 from n. Finally, we differentiate expr and when we have this result we copy it into the answer string after first placing down a $*$ sign. **Figure 4.2.1** shows the equivalent transformation on the original tree to produce the derivative tree. What EXPONENT needs to do is to mechanically turn one tree into the other.

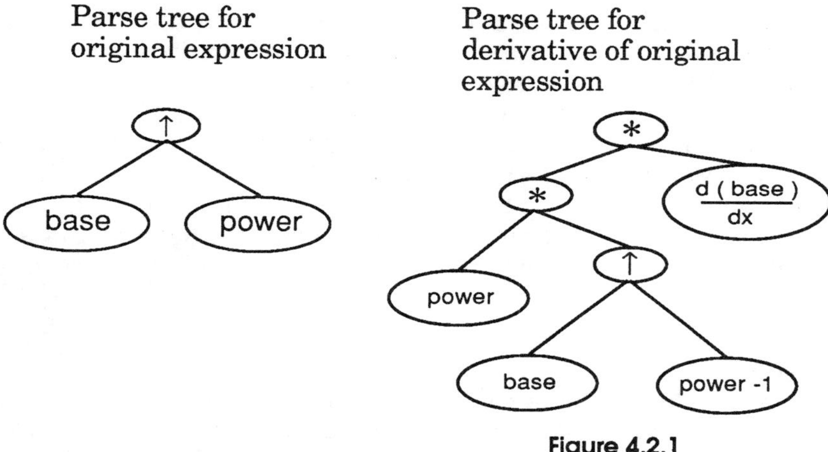

Figure 4.2.1

Here's how this tree manipulation could be done.

```
procedure EXPONENT ( input_tree: tree; var deriv_tree: tree);
(* declare various trees and sub-procedures*)

begin
    CHOP_LEFT(input_tree, base_tree);
    CHOP_RIGHT(input_tree, power_tree);
    Subtract 1 from contents of power_tree and make a
         new I-node tree with this integer as contents.
         Call this reduced_power_tree;
    CONNECT(base_tree, reduced_power_tree, '↑', tree_1);
    CONNECT(power_tree, tree_1, '*', tree_2);
    DERIVATIVE_OF_TREE(base_tree, deriv_base_tree);
    CONNECT(tree_2, deriv_base_tree, '*', deriv_tree)
end;
```

Some final comments

You may have noticed that the recursion in this process of taking derivatives is a bit different than the recursion we have seen in other sections of this pamphlet. In all our earlier examples, we had a recursive procedure which invoked itself. But DERIVATIVE_OF_TREE does not invoke itself. Instead, when we do not have a non-recursive exit, DERIVATIVE_OF_TREE invokes the appropriate subprocedure, based on the form of the tree, and that subprocedure is what reinvokes DERIVATIVE_OF_TREE. Thus we still get the effect of DERIVATIVE_OF_TREE being invoked from within itself (during execution of itself) – we still have recursion. This particular form of recursion is sometimes called a recursive chain.

Question 4.2.1

Can you write the procedures described here and thus create a program that does symbolic differentiation? You may need to learn more about trees from a book on data structures. such as Kruse [1984] or Tenenbaum and Augenstein [1986].

There are a number of ways in which the outline we have given here can be made into a more ambitious project. Programs like MACSYMA, MAPLE and MATHEMATICA don't need the excessive parentheses that result from strict adherence to our syntax diagram – they can handle expressions just as people write them. In addition they can deal with trigonometric functions, logarithms, the exponential function and other functions that our syntax diagram rules out. Furthermore, they can carry out algebraic simplification of expressions. The questions below address some of these issues.

Question 4.2.2

We have assumed that our algebraic expressions are heavily parenthesized so that they could easily be parsed into trees. If we want to be able to deal with expressions as they are normally written by people doing pencil and paper work, we will need to replace the syntax diagram we are using for an expression by something that allows the construction of expressions as human beings usually write them. This means that optional parentheses may or may not be present. The challenge here is not only to think up the appropriate syntax diagram, but also to write the routine PARSE. The version of PARSE needed to cope with our heavily parenthesized expressions is actually fairly easy. The parser for more "natural" expressions will be harder. Can you see how to construct this parser?

Question 4.2.3

We have left out the task of coping with trigonometric functions, the exponential function and the natural log function. Can you expand the program to include these functions?

Question 4.2.4

The distributive law asserts that $ab + ac = a*(b + c)$. For example, this can be applied to the expression $((x)*((3.1)+(x))) + ((x)*((x) - (2)))$ to yield this alternative but equal form: $(x) * (((3.1)+(x)) + ((x) - (2)))$. Can you write a program that accepts as input a parse tree for an expression and then checks whether the distributive law is applicable, and, if it is applicable, outputs a parse tree for the alternative form?

Section 4.3
Backward Chaining in Artificial Intelligence

In recent years, a field called Artificial Intelligence has arisen whose goal is to try to have computer programs solve problems which are usually regarded as requiring intelligence when human beings solve them. Although the field is quite young and still faces great challenges, impressive results have been obtained in some areas such as rule-based expert systems. These software systems make deductions in fields like medical diagnosis, oil exploration and equipment repair and maintenance. Rule based expert systems are undoubtedly one of the most practical advances in artificial intelligence.

Rule Based Systems

Table 4.3.1 shows a simple example of a rule-based system. This one is designed to give a student advice, in the form of rules, in the event of illness on the day of a dance she had been planning to attend.

rule	antecedent(s)	consequence
1	if you are infectious	then stay home from dance
2	if you won't be able to dance well *and* you won't disappoint anyone	then stay home from dance
3	if green spots on forehead	then stay home from dance
4	if green spots on forehead	then see doctor immediately
5	if cough *and* fever	then you are infectious
6	if headache and stomachache	then you won't be able to dance well
7	if you don't have a date	then you won't disappoint anyone by not going to the dance
8	if your date is a good sport	then you won't disappoint anyone by not going to the dance

Table 4.3.1

As you can see, each rule has an "if-part" consisting of one or more conditions called "antecedents" of the rule and a "then-part" which always consists of one condition called the "consequence." For example, Rule 2 has two antecedents: "you won't be able to dance well" and "you won't disappoint anyone."

Rule 2 has one consequence: "stay home from the dance". For a rule which has just one antecedent, if this antecedent is known to be true, then the consequence is true. If there is more than one antecedent, then if all are true we can deduce that the then-part is true. In terms of logical operations, we say that the truth value of the if-part consists of all the antecedent conditions "anded together." When we can perform a deduction of a then-part of a rule from the if-part, which we can do when the if-part is true, we say the rule *fires*. Notice that there can be many rules with the same then-part. For example, Rules 1, 2, and 3 have the same then-part. It is not necessary for all of these rules to fire in order for us to deduce the then-part. Either Rule 1 or Rule 2 or Rule 3 firing will be adequate to deduce that one should stay home from the dance.

AND/OR trees

A system of rules can be depicted in a diagram called an "AND/OR tree." In such a tree, the nodes are used to represent conditions. **Figure 4.3.1** shows the AND/OR tree for our example.

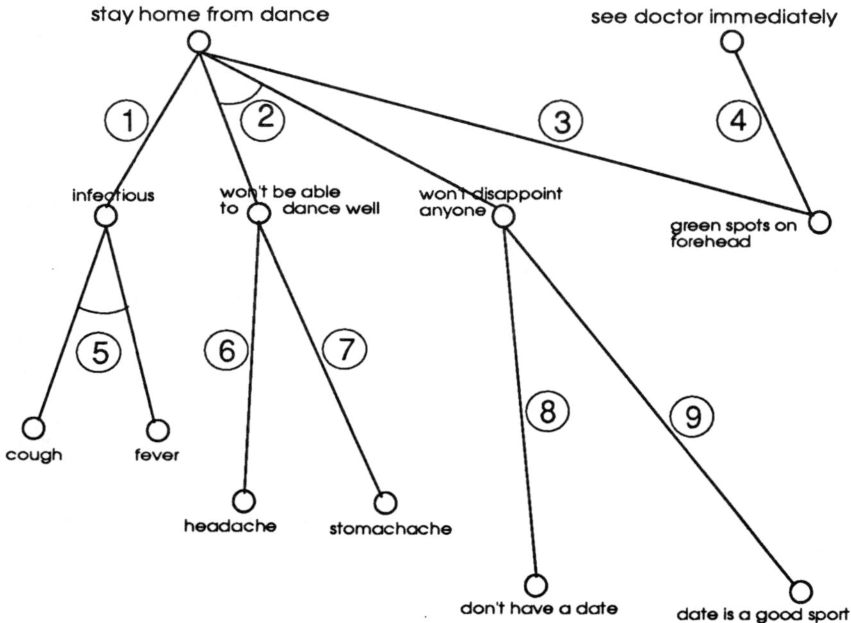

Figure 4.3.1

Conditions fall into two categories, observable conditions, like "don't have a date" whose truth we can easily find out and deducible conditions, like "infectious", whose truth (or falsity) must be deduced. Notice that the observable conditions are the ones which have nothing beneath them in the tree. A rule is represented either as a single link or a set of links between conditions. For example, Rule 1 is represented in the diagram by the link labeled 1. Rule 2, however, has two antecedents, both of which must be true in order for the consequence (the "then" part) to be deducible from that rule. In that case, we add a circular arc to connect the appropriate links.

Making deductions

Now suppose we know which observable conditions are true and which are false. Based on this information, we could determine the truth value of all the deducible conditions. For example, if "cough" and "fever" are both true, then we can deduce that "infectious" is true. In this case, we say that Rule 6 "fires." If either "headache" or "stomachache" is true then "won't be able to dance well" can be deduced. In this way, we can propagate truth values up the tree till we get to the top level nodes (the nodes which have nothing lying above them in the tree). This is called forward chaining.

There are instances where forward chaining is not very efficient. Forward chaining yields the values of *all* the top level nodes. If one only wants the truth value of one top level node, then doing the work to find them all can be time consuming. For example, in **Figure 4.3.2**, if we only wanted to know the truth value of condition A, it is only the bold-faced portion of the graph which is relevant. However, forward chaining, starting at the bottom nodes will give the truth values of all nodes. It might be nice to have a way of determining which part of the graph is the relevant part corresponding to the particular top condition one wishes to confirm. Backward chaining is a method which starts from the given top node, whose truth value you want, and works down to just the relevant bottom nodes.

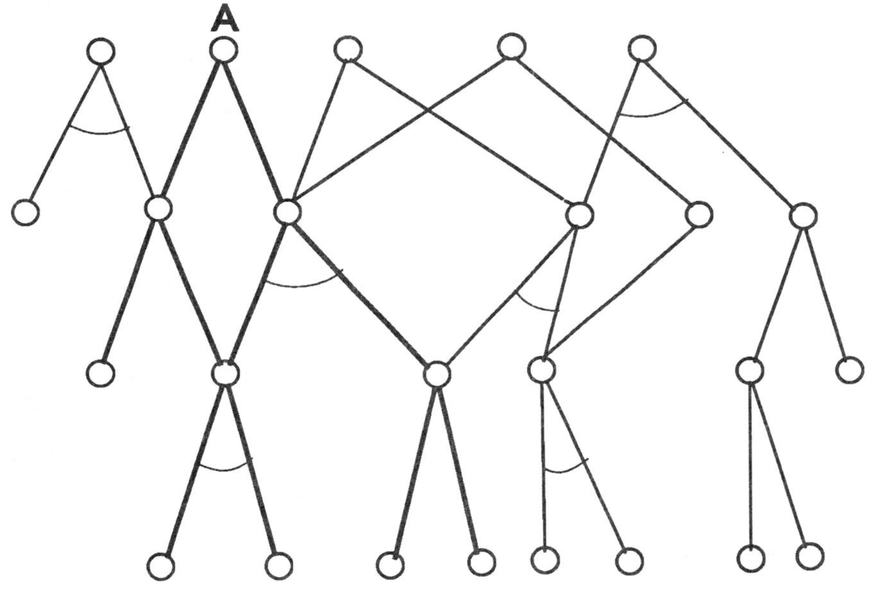

Figure 4.3.2

Let's see how backward chaining would work in **Figure 4.3.1.** Suppose we want to know whether to stay home from the dance. We will refer to this as attempting to evaluate the truth value of "stay home from the dance." First we ascertain that Rules 1, 2, and 3 are directly relevant in the sense that they all have "stay home from the dance" as their consequence. We call this set of rules the fan-in list of the condition "stay home from the dance." If any rule in the fan-in list fires, then the condition "stay home from the dance" must be true, since when the if-part of a rule is true, the then-part of that rule is also true.

To see whether Rule 1 fires, we need to know whether its one antecedent, the condition "infectious," is true or not. This is the same kind of problem we started out with, namely, determining the truth value of a condition, so we can evaluate antecedent conditions by recursive calls. If it turns out that "infectious" is true, then Rule 1 fires; since this firing means that "stay home from the dance" is true, we have our answer and can stop processing. If it turns out that "infectious" is false, then we go on to evaluate the next rule in the fan-in list, Rule 2. Because the if-part of Rule 2 has a two antecedents logically anded together, we must be prepared to evaluate both "won't be able to dance well" and "won't disappoint anyone." If either antecedent is false, then the entire if-part is false. Thus, as soon as one false antecedent is found, we can stop processing, since then we know that the entire if-part is false. Again, recursive calls are used to do the evaluation. If Rule 2 doesn't fire we go on to another recursive call to evaluate "green spots on forehead," the sole antecedent of Rule 3. Thus, the basic structure of the algorithm is a loop that checks through the fan-in list till it finds a rule that fires or until it runs out of rules to evaluate. If the algorithm processes the entire fan-in list without finding a rule that fires, then the algorithm reports that "stay home from the dance" is false.

As we have seen, recursive algorithms need non-recursive exits. In this problem, what this means is that we need eventually to come to a condition whose truth value we can evaluate without a recursion. This happens when we get to an observable node, whose truth value we can get by checking an array of known values.

The pseudocode below shows the basic structure of the algorithm EVALUATE, a function which evaluates and returns the truth value of a condition. In this pseudocode, we assume that the observable conditions are collected together in a set called "observable" which is global to this procedure. The true or false values for the observable conditions are stored in the global array called tfvalue. All other information required (for example, information needed to find the fan-in list) is also assumed to be available as global data. We leave to the reader the details of what local variables might be needed to carry out the loop. The function RULE_FIRES returns a boolean value (true or false) telling whether the rule it was sent fires or not.

```
function EVALUATE (node_to_check): boolean;

var it_fires  : boolean;
    tfvalue  : array[1 .. maxnodes] of boolean;

function RULE_FIRES(rule from fan_in_list): boolean;
(* code for this function shown in another example *)


begin (* EVALUATE *)
    if node_to_check is in the set "observable"
       then
          EVALUATE : = tfvalue[node_to_check]
       else
          begin
            ( * check rules in  fan_in_list of
               node_to_check *)
            repeat
              it_fires : = RULE_FIRES(rule from fan_in_list);
            until (it_fires) or
                  (no more rules in fan_in_list) ;
            EVALUATE : = it_fires
          end
end;
```

To check if a rule fires, we call upon the function RULE_FIRES which contains a loop to accommodate the fact that some rules, such as Rules 2 and 5, have more than one antecedent. In such a case, we need to evaluate each of the antecedents of the rule. This additional evaluating is accomplished by recursive calls to evaluate, of course. The loop starts by assuming that the rule will fire – a boolean variable "does_it_fire" is initially set to true – and then it loops through the antecedents until one is found that is false or until all antecedents are checked. If one is found to be false, then the rule doesn't fire. If the loop gets through the whole set of antecedents and finds each to be true, then the rule in question fires. Although this looping isn't needed for rules with just one antecedent condition, it is simpler to use a loop for these rules as well. Here is pseudocode for RULE_FIRES.

```
function RULE_FIRES(rule from fan_in list): boolean;

var does_it_fire : boolean;
    next_antecedent : condition_type;

begin
    does_it_fire := true;
    (* check if value of antecedents of this rule *)
    repeat
        determine next_antecedent;
        does_it_fire := (does_it_fire) and ( EVALUATE(next_antecedent) );
    until ( (all antecedents  checked) or (does_it_fire = false) );
     RULE_FIRES := does_it_fire
end;
```

The recursion in this algorithm is not direct, in the sense that a routine invokes itself, but takes the form of a recursive chain. In this recursive chain, one routine (EVALUATE) invokes another (RULE_FIRES) and that second routine invokes the first (EVALUATE). We saw another example of a recursive chain in **Section 4.2**. Recursive chains can also involve more than two routines.

Chapter 6 of Winston [1984] contains an elementary and readable account of rule-based expert systems and how they make use of backward chaining.

Question 4.3.1

Can you write the code to implement a rule-based system? The specific conditions, observable set, their values, and rules should be data for such a program, so that the program could be used to make deductions in various subject areas.

Chapter Five
About Recursion and Algorithms

I N **Chapter Three** we learned about recursive algorithms. Some of those algorithms evaluated functions defined recursively and some algorithms accomplished other ends such as sorting. In **Chapter Four** we saw several more complex applications of recursion. Hopefully by now we are pretty comfortable with recursion. Now it is time to delve more deeply into the hows and whys, or why nots, of expressing an algorithm recursively.

Also in this chapter we take a brief look at the proof technique known as mathematical induction and we introduce methods for evaluating the time complexity, or efficiency, of an algorithm.

Section 5.1
To Recurse or Not to Recurse

Let us first review and extend our ideas of the power of the recursive approach to an algorithm. It is instructive to look at the way in which we developed the MERGE_SORT algorithm. We took a top-down approach: We would have the whole list sorted if each half were sorted and we had a routine available to merge the two halves. This approach led directly to an algorithm as soon as we recognized that each half of the list could be sorted by a recursive call to MERGE_SORT. We took a large problem, delineated what would be necessary to complete the problem at the last step, and recognized smaller but identical problems in the preparation needed to get us to the last step. As we remarked earlier, this approach avoids our getting bogged down in messy details and the algorithm almost seems to write itself. There is something almost magical about this recursive approach in that it appears to give us something for nothing – we write the desired algorithm in part by assuming that the algorithm exists and works.

Two comments should be made about this top-down recursive approach. The first is that it tends to produce algorithms which are in some sense natural, easily understood by humans, and which often have shorter code than the corresponding nonrecursive algorithm for the same process. These are strongly positive attributes and cannot be ignored.

Second, recursive algorithms can often be executed very quickly on computers with parallel processors. An example of such an algorithm is our recursive MERGE_SORT. Imagine that the computer has many processors which can work simultaneously, just as the phone company is set up to allow for many simultaneous phone conversations. Then the processor assigned the first execution of MERGE_SORT can call upon two different processors to sort the two halves of the list. Each of them can call upon two other processors to sort the "quarters" of the list, and so on. The saving of time which can be achieved by having processors working in parallel is quite significant when compared to how long it would take if one processor had to do all the work itself. This saving is the basis of the trend in computer hardware design to provide independent parallel processing capability. What we should note here is that the top-down recursive approach to an algorithm is an approach which often leads to an algorithm capable of making use of the power of a parallel processing computer.

But didn't we hint earlier that recursion was sometimes not the way to devise an algorithm? Yes, we did; let's see why by examining recursive function evaluation for possible inefficiencies. Essentially the same considerations are valid when the algorithm under consideration is a procedure instead of a function.

Comparing costs of recursion and non-recursion

Let us return to a function we studied earlier. It was defined using a recurrence equation plus initial conditions:

$$f(n) = 2f(n - 1) + 3$$
$$f(0) = 1.$$

If we wish to have a computer procedure that evaluates this function we might write it recursively like this:

```
function F(n: 0 .. maxint): posint;

begin
    if n = 0
      then F := 1
      else  F := 2*F(n-1) + 3
end;
```

This recursive algorithm arises directly from the definition. But we could also write the following equivalent code:

```
function F_TWO(n: 0 .. maxint): posint;

var count, temp_ans: 0 .. maxint;

begin
   temp_ans := 1;
   if n > 0
     then
        for count := 1 to n do
            temp_ans := 2*temp_ans + 3;
   F_TWO := temp_ans
end;
```

This algorithm is not recursive. Since it has a **for** loop, it is a typical counting or iterative algorithm.

We see immediately that the code for iterative algorithm is a bit longer. In a simple case such as this, it would perhaps be meaningless to discuss which code is easier to understand. Both involve looping as part of execution. In the iterative algorithm the loop is evident. In the recursive algorithm the looping effect is created by the recursive call. If it were possible to get the result directly, with no looping, much time would be saved. In this case it is possible to get a direct result because we can use the recurrence equation and initial condition to find a formula for the function f. When we did this in **Chapter One** we got $f(n) = 2^{n+2} - 3$. If we use this definition of f, we get this evaluation algorithm:

```
function F_THREE(n: 0 .. maxint): posint;

begin
    F_THREE : = 2**(n+2) - 3
end;
```

This is clearly the most efficient of the three algorithms because there is no loop. Incidentally, the use of the double asterisk (**) to indicate exponentiation is not standard in Pascal. Many versions of Pascal do not support an exponentiation symbol; most other computer languages do have such a symbol.

Not all functions defined via a recurrence equation can be solved in terms of elementary functions and operations; one such function is n factorial. Evaluation in these cases requires that we choose between a recursive algorithm and an iterative one. Since both of these involve looping in some form, we must find some other characteristic in which they differ. At first glance it seems that the iterative algorithm uses more storage; it does have more variables. But this is deceptive. In all but the most trivial of cases, the recursive algorithm uses more storage and, usually, more execution time. Let us see why this is so using the example of computing the factorial function.

Here is a recursive computer function for evaluating $n!$:

```
function FACTORIAL(n: 0 .. maxint): posint;

begin
    if n = 0
        then FACTORIAL : = 1
        else FACTORIAL : = n*FACTORIAL(n-1)
end;
```

Suppose this function has been invoked and sent the value of 7 for n. As execution proceeds on the **else** branch, the multiplication called for cannot yet be performed because although the value of n is known ($n = 7$), the value of FACTORIAL $(n - 1)$ is not yet known. Before the recursive call to FACTORIAL is made, sending it a 6 (since $n - 1 = 6$), something must be done (by the computer, not by the programmer) to ensure that when we return to this multiplication operation later, we have both the value of FACTORIAL(6) and the value of 7 for n. The current value of n, namely 7, must be saved for later use. Similarly, during the execution of the function to determine FACTORIAL(6), the then current value of n, now a 6, must be saved. Actually, all values of n between 7 and 1 wind up being saved before we invoke FACTORIAL with an n of 0 and branch to the non-recursive exit.

Since it is necessary in many recursive algorithms to save current values of variables for future use, that is, use after the return is made from the recursive call, storage is needed to implement these recursive calls. This storage is not declared by the program but is created and maintained by the Pascal system. As we can see from this example, for even small values of n the amount of this "hidden" storage needed is greater than the obvious "extra" storage for two variables used in the following iterative version of an n factorial function:

```
function FACTORIAL_TWO(n: 0 .. maxint): posint;

var ans, count: 0 .. maxint;

begin
    ans : = 1;
    if n > 0
      then
          for count : = 1 to n do ans : = count * ans;
      FACTORIAL_TWO : = ans
  end;
```

The management of the "hidden" storage by the Pascal system is an interesting topic which we shall briefly touch upon here. In our example the values of n in the order in which they are stored are 7, 6, 5, 4, 3, 2, 1. But the order in which those values are later used is exactly reverse from the storage order. Once FACTORIAL(0) of 1 is returned, the value of n is restored to 1 so that FACTORIAL(1) can be calculated. That value is than returned and the value of n is restored to 2 so that FACTORIAL(2) can be calculated. And so on until FACTORIAL(7) is calculated and returned to the proper location external to the function. We see that the last value to be stored, the 1, is the first value to come out of storage and be used. An arrangement of data in this last-in-first-out (LIFO) manner is called a stack. Stacks are discussed in books on data structures such as Kruse [1984] and Tenenbaum and Augenstein [1986].

Question 5.1.1

Suppose we have an algebraic expression satisfying the syntax diagram in **Section 2.2**. Further suppose that it contains no variable names and only one digit positive numbers. Can you write an algorithm that accepts such an expression as a string of characters and determines the numeric value of the expression? For example, if "2 + 3*5" is the string then we would want the returned value to be 17. Note that the order of operations is crucial here. Solving this non-trivial problem involves the use of stacks. Hints can be found in many data structures texts such as Kruse [1984] and Tenenbaum and Augenstein [1986].

We are now in a better position to understand the overhead costs involved in recursion. Storage is needed for the stack of data. In addition to maintaining this stack, the Pascal system has other tasks to perform to ensure that the call to and return from any subprocedure or function is handled smoothly. Thus a recursive algorithm in a computer program typically incurs both space and time requirements in excess of the needs of a program using an equivalent non-recursive algorithm. So now we see a real dilemma: Is the clarity and elegance of recursion (for the human) worth the price of extra time and space in the computer?

All of this would be just an intellectual exercise were it not for the fact that there is a method which a programmer can use to turn any recursive routine into a non-recursive one. When this is done, the resulting code is always a mess (although it can be cleaned up), it often obscures the underlying algorithm, and it requires that the program itself maintain the stack of data. A description of the method and examples of it can be found in many books on data structures (Kruse, 1984). In practice, many algorithms are written recursively because recursion is a natural and easy way to approach many problems. Then, if a computer implementation is to be used frequently, and if translating the algorithm to a non-recursive equivalent would greatly reduce the time and storage required for execution, the translation is made.

Sometimes recursion is clearly the wrong way to program

In some cases, the recursive algorithm is grossly inefficient when compared to the non-recursive one because its execution involves the redoing of work that has already been done. Such a case exists in evaluating the Fibonacci function which we met in **Chapter One**. To demonstrate this, we compare two programs to evaluate that function. First we have the non-recursive one:

```
function FIBON_LOOP(n: posint): posint;

var first, second, total, count: posint;

begin
   if (n = 1) or (n = 2)
     then FIBON_LOOP := 1
     else
       begin
         first := 1;
         second := 1;
         for count := 3 to n do
           begin
             total := first + second;
             first := second;
             second := total
           end;
         FIBON_LOOP := total
       end
 end;
```

We see that in this iterative algorithm, if $n > 2$, the value of the function is calculated by calculating the value for $n = 3$, then the value for $n = 4$, and so on until the desired value of the function is reached.

Next we look at a recursive algorithm which accomplishes the same task:

```
function FIBON_RECUR(n: posint): posint;

begin
   if (n = 1) or (n = 2)
      then FIBON_RECUR : = 1
      else FIBON_RECUR : = FIBON_RECUR(n-1) + FIBON_RECUR(n-2)
end;
```

Consider what happens when this recursive version is invoked with 6 as the value of n. On the else branch we see recursive calls for parameter values of 5 and 4. So far so good. But the execution of the call with value 5 will result in calls with values of 4 and 3. There will be two identical calls with a value of 4 and the second one of these to execute will not "know" that the work has already been done, so each will execute fully (see **Figure 5.1.1**). In addition, each call with a value of 4 will generate a call with a value of 3. But we already had such a call from the call with a value of 5. So the function will be called upon three times to find FIBON_RECUR(3). This situation gets worse and worse as the incoming value of n on a call gets smaller and farther away from the original value of n on the first, non-recursive call to the function. (See **Question 5.1.2**). There is no contest between the recursive and non-recursive algorithms here: The non-recursive version wins the efficiency race hands down.

Question 5.1.2

Can you determine how many calls there are in FIBON_RECUR with parameter values of 3, 2, and 1 in the case that the original call was with n of 7? Can you follow the same analysis at least part way for an original call with a value of 31 for n? Do you see a pattern in the number of recursive calls for a particular value of n and the distance between that value and the original value of n?

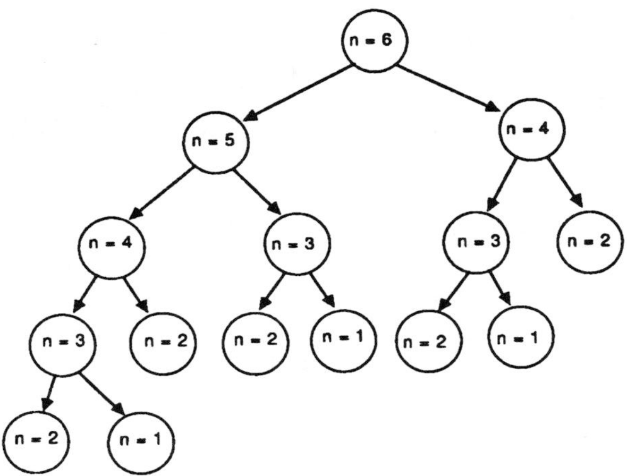

Figure 5.1.1 The pattern of recursive calls during execution of the function FIBON_RECUR.

Section 5.2
Recursion and Induction

It has been said jokingly that mathematicians never produce any new results – they simply reduce new problems to variations of problems they have already solved. For example, to divide by a fraction – a "new" problem – invert the fraction and multiply – a process already understood. In algebra, solving a third degree equation is usually approached by finding one root and reducing the problem to finding the roots of a quadratic, again an already understood process. It would appear that computer scientists go the mathematicians one better. They seem to solve large complex problems by assuming that the exact same problem is soluble on a smaller scale. The key ingredient in recursion is being able to specify exactly how the solution to the smaller problem(s) can be used to solve the larger one. The knowledge of how to get from one stage to the next is crucial to the recursion.

Mathematicians have been using an analogous thought process for years for proving certain kinds of theorems. This proof method is called *mathematical induction*. It is typically used to prove theorems that state that something is true for infinitely many integers. As an example, consider the statement that:

(5.2.1) $1 + 2 + 3 + \ldots + n = n(n + 1)/2$ for every positive integer n.

How would we convince ourselves that this is true? We could easily substitute a particular value of n to test the formula for a single case. For example, if we substitute $n = 1$, we obtain

$$1 = 1(2)/2$$

which is true. But this only verifies the theorem for one value. What about $n = 2$, $n = 3$, $n = 4$, etc? The crucial step in the proof by induction is showing that if one assumes the theorem to be true for some value of n, say $n = k$, then we can show that the theorem is true for $n = k + 1$. That is, if we know the result at one stage, we use it to deduce the result at the next stage.

Here's how this would work in the case of formula (5.2.1). Suppose we knew the formula were true for a particular k, i. e.

$$1 + 2 + \ldots + k = k(k + 1)/2$$

Then it follows that

$1 + 2 + \ldots + k + (k + 1)$

$$= (1 + 2 + \ldots + k) + (k + 1) \quad \text{(insert parentheses)}$$

$$= k(k + 1)/2 + (k + 1) \quad \text{(substitute formula for } k\text{)}$$

$$= (k + 1)(k/2 + 1) \quad \text{(algebra)}$$

$$= (k + 1)(k + 2)/2 \quad \text{(more algebra)}$$

The last formula is exactly what we would get if we substitute $k + 1$ for n into formula (5.2.1) which we wish to prove. Thus we have shown that anytime the formula is true for one integer, it is true for the next. We have already seen that the formula is true for $n = 1$. Consequently, the formula holds for all positive integers.

Mathematical induction is closely related to the recurrence equations used to describe functions $f(n)$ which we explored in the first part of this pamphlet. Induction might be used to prove the formulas we developed for those functions. The mathematician usually uses induction to expand the scope of a result, to go from k to $k + 1$. The computer scientist uses similar reasoning from the opposite point of view: A problem of large scope is solved by relating it to identical ones of smaller scope.

The method of proof by induction also has a part that corresponds to the non-recursive exit in a recursive algorithm. In a proof by induction we must show that there is some starting value of n for which the theorem holds true. In **Section 1.2** when we verified that a formula for a function satisfied a recurrence relation plus initial condition, our work was similar to a proof by induction. Let us look at such an exercise once more, this time as a proof by induction:

Example

Show that $f(n) = 2^{n+2} - 3$ *satisfies this recurrence relation plus initial condition:*

$$f(n) = 2f(n-1) + 3 \qquad n = 1, 2, 3, \ldots$$
$$f(0) = 1.$$

We begin by showing that for some initial value of n *the formula holds; for obvious reasons we choose* n = 0. *Substitution yields* $2^{0+2} - 3$ *or* $4 - 3$ *or* 1, *which is the correct value.*

Now let's see how the inductive step works. We assume that the formula is correct for f(k) *and show that under this assumption the formula is also correct for* f(k + 1).

$$f(k+1) = 2f(k) + 3 \qquad \textit{(recurrence relation)}$$

$$= 2(2^{k+2} - 3) + 3 \qquad \textit{(use assumption)}$$

$$= 2^{k+3} - 6 + 3 \qquad \textit{(algebra)}$$

$$= 2^{(k+1)+2} - 3 \qquad \textit{(more algebra)}$$

And on the last line we have the appropriate formula for the case of k + 1. *Hence, we can conclude that* f(n) $= 2^{n+2} - 3$ *for all non-negative integers.*

Proofs by induction are quite common in mathematics whenever it is desired to show that a result holds for all positive integers. You can read more about proof by induction in many mathematics texts, especially in discrete mathematics texts such as Kolman and Busby [1984].

Question 5.2.1

Can you rewrite the other verifications of formulas for recurrence equations from **Section 1.2** as proofs by induction? Of course, the general form of the theorem you are trying to prove is: This formula for $f(n)$ satisfies this recurrence equation plus initial condition.

In **Section 1.2** we proved formulas for arithmetic and geometric sums. Those formulas are often proven using induction. Here we outline as questions proofs for two of those formulas.

Question 5.2.2

To show that the sum of the first n terms of the arithmetic series having first term a and common difference of d is $n(2a + (n-1)d)/2$ we would start with the initial step of showing the formula true for $n = 1$. The series would have just one term, namely a, and thus would sum to a. Substituting 1 for n into the formula also gives a, so we have our initial step. In the inductive step, we assume that the formula is true for $n = k$ and try to show its truth for $n = k + 1$. We start by writing out the $k + 1$ terms:

$$a + (a + d) + (a + 2d) + \ldots + (a + kd) + (a + (k + 1)d)$$

and then we insert parentheses:

$$[a + (a + d) + (a + 2d) + \ldots + (a + kd)] + [a + (k + 1)d]$$

Can you substitute the correct formula for the first bracketed expression, the sum of the first k terms, and then do the algebra to complete the proof?

Question 5.2.3

In a manner similar to the one outlined in the previous question, try to verify by induction that the sum of the first n terms of a geometric series starting with the value a and having a common ratio of r is $a(r^n - 1)(r - 1)$.

Section 5.3
Efficiency of algorithms

A very important criterion used to evaluate different algorithms which perform the same task, such as sorting, is the efficiency of the algorithm. In measuring the efficiency, we try to determine, in terms of the number of pieces of data to be processed, how much work the algorithm actually does. This measure of work done can then be used to judge which algorithm will complete its work sooner, given that all algorithms being compared will be executed on the same computer under comparable circumstances.

In **Chapter Three**, when we looked at two methods of implementing phone chains, we noted that it took longer for everyone to get the message using PHONE_CHAIN_ONE than using PHONE_CHAIN_TWO. This difference is directly attributable to the fact that in PHONE_CHAIN_TWO, simultaneous phone calls were being made, but phone calls were not made simultaneously in PHONE_CHAIN_ONE. We should note that the number of telephone calls made in each phone chain is exactly the same. In our example, the volleyball team had 15 members. After the first person received the message, each of the other people eventually also was given the message. Thus each phone chain resulted in 14 phone calls being made. If simultaneous phone calls were not possible, there would be no savings of time in choosing PHONE_CHAIN_TWO. This is important for us to understand because often in computer applications, the computer being used is one in which, effectively, only one processor is available for executing an algorithm. This is equivalent to allowing only one phone call at a time. The kind of comparison we are looking for among algorithms will be based on the amount of work to be done, under the assumption that only one processor is available.

Efficiency of linear search

Here is an example of the kind of analysis we do to determine efficiency. Let us look at the linear search from **Chapter Three**. Suppose we are searching for an item in an existing array of n entries. The basic unit of work done in a search is a comparison of two items – the search key with the corresponding field of the array entry we are examining. We need to estimate the average number of comparisons in a linear search. If the desired item is the first entry, we do one comparison; by contrast, if the desired item is not in the array, we do n comparisons before we discover that fact. Since even if the desired item does exist in the array, we could do anywhere from 1 to n comparisons, we need to find the average number of comparisons. Let's assume that the element we are looking for is equally likely to be in any of the n positions. Then the number of comparisons we need to do is equally likely to be 1, 2, . . ., n.. Thus the average number of comparisons is:

$$(1 + 2 + 3 + 4 + \ldots + n)/n$$

which has as its numerator the sum of the first n positive integers. The numerator can be summed using the formula for an arithmetic series, yielding $n(n + 1)/2$. So the average we want is $(n + 1)/2$.

Theorem 1

The average number of comparisons required by linear search acting on a list of n items is $(n + 1)/2$.

For reasons of simplicity, and because for large values of n the error is negligible, we often say that for a linear search through an array of n items, the average number of comparisons is $n/2$.

If n is a "small" number and we do not do many searches, then we can live with linear search very nicely. But if the array has 8000 entries and it is searched 500 times a day, then the average amount of daily work done is considerable: $500(8000/2)$ = 2,000,000 comparisons. We may not be able to change the fact that we need to do 500 searches. But we surely should investigate whether the 8000/2 is the best average efficiency we can achieve for each individual search. In fact more efficient search techniques do exist if the array is sorted. The binary search technique described in **Question 3.2.6** is more efficient.

Question 5.3.1
Can you see why binary search is more efficient than linear search?

Efficiency of some sorting algorithms

A common unit of work considered in a sort algorithm is the moving or switching of array entries. In the insertion sort (**Section 3.2**), we wound up inserting the items one at a time into an ever growing subarray of already sorted items. Except in the trivial case of the new item being the first position of the array, we move the new item twice: first into a temporary location and then into its proper position in the array once that position is found. The location for the new item is found by a reverse order linear search of the sorted subarray. If we are inserting the 15th item, we do a linear search of the sorted subarray of 14 items and on the average make 7 comparisons before finding where to do the insertion. And, more to the point, we must move down an average of 7 items to make the room for the insertion. Thus, to insert a particular item, the tally of moves required on average is: two for the item to insert -- plus half the number of already sorted items to make room for the new item. A table will help us find the average number of moves in an insertion sort on an array of n items:

inserting item number	size of sorted subarray	average number of moves
1	0	0
2	1	$1/2 + 2$
3	2	$2/2 + 2$
4	3	$3/2 + 2$
5	4	$4/2 + 2$
.	.	.
.	.	.
.	.	.
n	$n - 1$	$(n - 1)/2 + 2$

Average number of moves in INSERT_SORT

Table 5.3.1

To find the average total number of moves for the whole sort, we add up the averages in the third column, separating out the varying numerical part on the left from the " + 2" part:

$$(0 + 1/2 + 2/2 + 3/2 + 4/2 + 5/2 + \ldots + (n - 1)/2) + 2(n - 1)$$

The parenthesized sum on the left is half the sum of $1 + 2 + 3 + 4 + 5 + \ldots + (n - 1)$ which is the sum of the first $n - 1$ integers, an arithmetic series. The standard summation formula tells us that the sum of the first $n - 1$ integers is $n(n - 1)/2$. So our parenthesized sum is $n(n - 1)/4$ and the desired formula is $n(n - 1)/4 + 2(n - 1)$ or $(n^2 + 7n - 8)/4$. Thus, we have proved:

Theorem 2

The average number of data moves required by INSERT_SORT when the input list contains n items to be sorted is $(n^2 + 7n - 8)/4$.

For example, if we use insertion sort on an array of 400 items, then on the average there would be $(400^2 + (7)(400) - 8)/4$ or 40698 data moves.

Question 5.3.2

Would our formula $(n^2 + 7n - 8)/4$ for the average number of data moves in an insertion sort of n items be different if a different search technique were used to locate the insertion position for each item?

Question 5.3.3

Suppose we decided to modify the insert procedure so that it only moves the item to be inserted into a temporary location if the item is not already in the correct location. In other words, if the item to be inserted is larger than all those in the sorted portion of the array, no data moves are made at all. Can you find a formula for the average number of data moves made using this modified version in an insertion sort of n items? The formula should evaluate to less than the one in the theorem. *Hint*: the probability that an item is larger than k other items is $1/(k - 1)$.

We now investigate how much work is done by the procedure MERGE_SORT, again with an array of n items. For convenience, since in MERGE_SORT the array is split physically into halves, which then are in turn split into halves, we choose n so that it is a power of 2. That is $n = 2^m$ for some integer m; and consequently, $m = \log_2(n)$. In our procedure MERGE_SORT, the actual moving of data is done by the subprocedure MERGE. When MERGE is sent two subarrays of length k, it moves each of the $2k$ elements into the array temp and then back into the proper position in data_array. So there are $4k$ data moves (into and out of the array temp) in a MERGE of two subarrays of length k. Due to the recursion in MERGE_SORT, MERGE is called a number of times with different k's. The table below helps us see how many times MERGE is sent subarrays of different lengths and how much total work is done by MERGE.

length of subarray (k)	no. of times two subarrays of length k are merged (t)	work done ($4kt$)
$n/2 = 2^{m-1}$	$1 = 2^0$	$4(n/2)(1) = 2n$
$n/4 = 2^{m-2}$	$2 = 2^1$	$4(n/4)(2) = 2n$
$n/8 = 2^{m-3}$	$4 = 2^2$	$4(n/8)(4) = 2n$
.	.	.
.	.	.
.	.	.
$n/2^m = 2^{m-m}$	$n/2 = 2^{m-1}$	$4(1)(n/2) = 2n$

Number of data moves done by MERGE_SORT

Table 5.3.2

Let's check the validity of the entries on the bottom row of **Table 5.3.2**. First we look at the left column and note that since we chose $n = 2^m$, the expression at the bottom of the first column must be equal to 1. But subarrays of length 1 are the shortest ones merged in the merge sort, so we have the right expression at the bottom of the left column. Next we look at the bottom entry in the middle column. Consider how many pairs of adjacent single-element lists are there in an array of n elements? There are $n/2$ pairs, so this expression is also correct. Since all of the entries in the right column are $2n$, the bottom entry in that column is not a surprise.

The total work done by MERGE is the sum of the entries in the right column. That sum would be easy to find if we knew the number of rows in the table. But we can see by the patterns which have been developed in the left and middle columns of the table that there are m rows in the table. Since $m = \log_2(n)$, the amount of work done in a merge sort is the product of $2n$ and $\log_2 n$. The foregoing analysis amounts to a proof of the following theorem:

Theorem 3

The number of data moves required by MERGE_SORT acting upon a list of n = 2^m *items to be sorted is, in every case, 2nlog2n.*

Question 5.3.4

Notice that the theorem does not speak of the average number of data moves. Can you convince yourself that, no matter what the data list is like, the number of moves will always be $2n\log_2 n$. In other words the best, worst and average work required is the same for MERGE_SORT.

The average case is not the only case

In our discussions of searches and sorts, we have looked for the average amount of work to be done. For some algorithms, the average case is the only case: This is the situation for a merge sort. But for many algorithms, the amount of work done varies with the arrangement of the data being processed. Consider, for example, the linear search. We found that for an

Page	(line)	Instead of	Read
vi	Fig. 20 caption		l.c. ital. *p*
3	−11	work	word
23	−10	like nim-heap	like a nim-heap
24	−9	1 x 1 x 2	$1 \times 1 \times 2$
38	−11	,denote by S{2,5,7}	(delete 2nd occurrence)
43	−12	of *b* ..., of *c* ...	or *b* ..., or *c* ...
47	2 in box	0	0
48	−2	connections	connexions
50	above Ex. 37	know	knows
51	−3	omitted)	respectively) (move rt. paren.)
	−2	than million	than a million
59	Fig. 24		(insert row of turtles)

Page	(line)	Instead of	Read
63	1 in box	heap up	head-up
70	6	A x B	$A \times B$
		(this page is part of Project F, and should be in italic?)	
71	−10	possible	possibly
78	2 above Fig.	$= (31 \; \ddagger \;) \; \ddagger \; 33 =$	$= (31 + 1) \; \ddagger \; 33 =$
80	−2	$= (b + c) = 1$	$= (b + c) + 1$
94	−4	theheaps	the heaps
95	−11	th	the
102	2 of 54.	more	move
	8 of 54.		(omit second comma)
104	3 of frieze	... 12 22 48	... 12 24 8
105	3 of 6.	numbers	nimbers
	8 of 6.	Hackbush	Hackenbush
106	10.	1934. *F.C.R.*	*F.C.R.*, Dec. 1934,
107	32.	Tydschrift	Tijdschrift
108	37.	Tikoku	Tôhoku
	39.	Mathematica	Mathematics
	40.	te orji	teorji
110		Hickery	Hickory
	mating	$= (a \; b) - 1$	$= (a \; \ddagger \; b) - 1$
111		n-position	$\mathcal{N}$-position
		O-position	$\mathcal{O}$-position
112	open	O-positions	$\mathcal{O}$-positions
		P-position	$\mathcal{P}$-position
	Sivler	Sivler	Silver
113	Turning Tables	59-60, 42	59-60, 62-64
	Welter function	Welter's Game	(move onto next line)
	Welter's Game		71-73

average case, a linear search needed to search $(n + 1)/2$ items before finding a match. In the worst case, we would have to search all n items.

In a linear search, the number of "wrong" items which must be processed before a match is found is crucial to the amount of work done. There is one circumstance in which that number can be effectively minimized: if you had a pretty good idea of where in the list the search key might be found. If there is certain small subset of the items in the array for which there is a high probability that the search key is in that subset, then simply by bringing all items of high probability to the "front" of the array, we shorten the search.

Question 5.3.5

Suppose that an array of 10000 items is arranged so that the 100 items which make up 90 percent of the searches are the first 100 items. What is the average number of searches needed to locate an item in this array? (*Hint*: Use a weighted average.)

Question 5.3.6

The "bring-to-the-front" technique for maintaining an array for linear searching works like this: whenever you find an item you have been searching for, put that item in the first position of the array. This is done either by switching the items in the first and the newly matched position or by moving down the items in between as was done in the insertion sort. Do these methods accomplish the task of bringing frequently needed items to the front of the array? Is there a difference in the efficiency of these two methods?

Each sort algorithm differs in what arrangement of data would constitute a worst case, an average case, or a best case. In the case of an insertion sort, the best case — the one involving the least amount of work — is one in which there is never any need to shift data down during an insertion. This means that each new piece of data is larger than the data already processed. This best case occurs when the data in the array is already in order. The other side of the coin would be a data arrangement in which every insertion involved moving down every piece of data in the sorted subarray. This would occur if each new piece of data were smaller than all the items already processed; the worst case in an insertion sort is data already sorted, but in reverse order.

Question 5.3.7

We used data moves to measure the work done in an insertion sort, but in the best case scenario with the modified version of insert described in **Question 5.3.3**, no moves are done. Would you say that no work is done in this case? What basic measure might you substitute for data moves? Is it "fair" to compare two algorithms if the basic work element is not the same in each case?

Theoretical considerations:
Measuring work using recurrence relations

Since we began this pamphlet by discussing recurrence equations, then applied recursive programming techniques to recurrence equations, it seems fitting to ask whether recurrence equations can somehow be applied to analyzing recursive programs. We conclude this pamphlet by showing how we can use recurrence equations to get information about the amount of work done by a recursive algorithm in the worst case. Although we will illustrate the method with sorting algorithms, the method is very general and can be applied to any recursive algorithm.

For a given algorithm, we define a work function w like this: $w(n)$ is the amount of work done if there are n pieces of data sent to the algorithm. We will find that having written our algorithms recursively helps us see what recurrence equation is appropriate.

The principle we will apply can be explained like this. Suppose an algorithm has two parts and the worst amount of work ever required for part A is w_A, while the worst amount of work ever required for part B is w_B. Then the worst amount of work, w, for the whole algorithm must satisfy:

$$w \le w_A + w_B$$

Note that we cannot assert that we always have equality, $w = w_A + w_B$, even though there may be some algorithms for which equality holds. Here is an example from everyday life that illustrates the issue. Suppose that mowing the lawn requires two steps: preparation of the mower, which might include filling it with gas, sharpening the blade, etc.; and using the mower on the grass. The worst case for preparation is 1 hour and this occurs when sharpening is needed. The worst case for using the mower is 1 hour and this occurs when the blade is dull. However, the worst case for the whole job is never 2 hours since the two worst case situations never occur at the same time.

Now suppose we had a recursive algorithm which has the following general form:

```
procedure F(n: integer; other data);

begin
    F(n-1, other data);
    non-recursive part
end;
```

The recursive and non-recursive parts are like part A and part B in our previous discussion. Let $w(n)$ be the worst case number of operations for the entire procedure operating on n data items and let $v(n)$ be the worst case number of operations for the non-recursive part. Then

$$(5.3.1) \qquad w(n) \leq w(n-1) + v(n)$$

Normally, $v(n)$ would be some known function of n and w is the unknown function we would like to solve for.

If we had an "=" in place of the "≤" in (5.3.1) we might find $w(n)$ using the methods described in **Chapter One**. But the "≤" makes it impossible to get an exact formula for w in this way. What we can do instead is to provide a formula for an upper bound to w, i.e., we will find a formula for a function f with the characteristic that

$$w(n) \leq f(n) \text{ for all } n \geq 0$$

The f we want turns out to be the solution to the recurrence equation we obtain by simply changing the "≤" to "=". This is justified by the following theorem.

Theorem 4

Let w, f *and* v *be functions and let* a > 0. *Suppose:*

1. w(1) = f(1)
2. w(n) ≤ a * w(n - 1) + v(n) for n ≥ 2
3. f(n) = a * f(n - 1) + v(n) for n ≥ 2
Then w(n) ≤ f(n) *for all* n ≥ 1.

Question 5.3.8

We outline here a proof of **Theorem 4** using the technique of induction. Can you fill in the gaps?

1. Basic step: $n = 1$. Why is $w(1) \leq f(1)$?

2. Inductive step: We assume that $w(k) \leq f(k)$ for some specific integer k and use that assumption to show that $w(k + 1) \leq f(k + 1)$. Start by rewriting 2) of the theorem substituting $k + 1$ for n. Next substitute $f(k)$ for $w(k)$ on the right, noting that this is justified by our inductive hypothesis and does not disturb the inequality. By now, the expression on the right should match the one you would get by substituting $k + 1$ for n into 3) of the theorem, so you have shown that $w(k + 1) \leq f(k + 1)$.

Applying the theory to sorting algorithms

Now let's apply these ideas to INSERT_SORT. In INSERT_SORT, the amount of work (data moves) needed to sort a list consisting of one element is no work at all:

$$w(1) = 0.$$

In INSERT_SORT, we used a recursive call to insure that the first $n - 1$ items in the array were sorted, then as the non-recursive part, we invoked insert to position the nth item. The worst amount of work needed to insert the nth item occurs when each of the previous $n - 1$ items needs to be moved. In addition, we have to move the item to be inserted into a temporary location outside the array so that the $n - 1$ items can be moved, and then we have to move the item to be inserted from its temporary location to its proper place. Altogether, that makes $n + 1$ moves. So we have this recurrence equation for the amount of work (in the worst case) done up to and including the insertion of the nth item:

$$w(n) \leq w(n - 1) + (n + 1).$$

Following **Theorem 4**, we can get an upper bound on w by solving the related recurrence equation and initial condition
$$f(n) = f(n - 1) + (n + 1) \qquad n = 2, 3, 4, \ldots$$
$$f(1) = 0$$

We solved this equation earlier in **Section 1.2** where we obtained
$f(n) = (n^2 + 3n - 4)/2$. Since $w(n) \leq f(n)$ this proves:

Theorem 5

The worst number of moves required by INSERT_SORT *when acting on a list of* n *items to be sorted is no worse than* $(n^2 + 3n - 4)/2$.

We note that $(n^2 + 3n - 4)/2$ is not the formula we obtained when discussing INSERT_SORT earlier in this section. In our earlier discussion, we were analyzing the average number of moves, not the worst case, so it is not surprising that the formula we reached for the work done in the average case, $(n^2 + 7n - 8)/4$, shows less work being done in the average case than in the worst case.

Finally, let's analyze MERGE_SORT using recurrence equations. As we did earlier in discussing MERGE_SORT, let us assume that n is a power of 2. How much work is done if we use MERGE_SORT to sort an array of one item? Of course

$$w(1) = 0.$$

The worst amount of work (number of data moves) done to sort an array of n items is the amount of work needed to sort two arrays of $n/2$ items using MERGE_SORT (the two recursive calls) plus the amount of work done in the merge routine, which we recall is 4 times the length of the arrays being merged. So we have

$$w(n) \le 2w(n/2) + 4(n/2).$$

Inspired by **Theorem 4**, we try to solve the related recurrence equation and initial condition:

$$f(n) = 2f(n/2) + 2n \qquad n = 2, 3, 4, \ldots$$
$$f(1) = 0$$

This was solved in **Section 1.2**, where we got: $f(n) = 2n\log_2(n)$. You may have some difficulty spotting this calculation in **Section 1.2**, because there we had set $n = 2^m$ and so the equation became

$$f(2^{m+1}) = 2f(2^m) + 2^{m+2} \qquad m = 1, 2, 3, \ldots$$
$$f(2^0) = 0,$$

and found the solution of $f(2^m) = m2^{m+1}$. But since n has been chosen as 2^m, then $m = \log_2(n)$, and these are really the same equations.

Earlier in this section we analyzed MERGE_SORT another way, not involving recurrence equations, and discovered that the number of data moves needed is always exactly $2n\log_2 n$. This new calculation confirms that the best, worst and average cases of MERGE_SORT don't differ – there really is only one case as far as the number of data moves is concerned. Comparing this earlier analysis with the recurrence equation approach shows that, for this algorithm, the analysis based on recurrence equations gives less detailed information. However, there are other algorithms where the recurrence equation approach gives information which is not readily obtained in other ways.

Question 5.3.9

Can you write the appropriate recurrence equation plus initial condition for the worst case of a linear search? Can you solve the system?

Question 5.3.10

So far we have measured work done in a sort routine by the number of data moves required. We could instead use the number of comparisons between two items as the unit of work. Can you do worst and average case analyses for MERGE_SORT and INSERT_SORT using this unit of work?

Question 5.3.11

What is the ratio of worst-to-best case efficiency for INSERT_SORT? Can you show that this approaches 2 as n approaches infinity?

Question 5.3.12

There are a great many other sort algorithms discussed in data structures texts such as Kruse [1984] and Tenenbaum and Augenstein [1986]. Study some of these algorithms, such as heap-sort, bubble-sort, and selection-sort, in regard to the question of what the ratio typically is between worst and average case efficiencies.

Drawing conclusions based on efficiency considerations

The kind of efficiency comparisons and best vs. average vs. worst case considerations we have just been discussing are included in texts on data structures such as Tenenbaum and Augenstein [1986].

We have studied two sort algorithms, INSERT_SORT and MERGE_SORT. In the worst and average cases, they clearly have different amounts of work to do. Thus in the worst or average cases, for randomly arranged data, MERGE_SORT is a more efficient way to sort the data items. But we cannot make a blanket statement that MERGE_SORT is always to be preferred to INSERT_SORT. The best case for MERGE_SORT requires the same amount of work as both other cases, while INSERT_SORT with the modified version of insert requires no data moves if the data is already ordered.

You can imagine that with dozens of sorting algorithms having been developed, the choice of which algorithm to use is not a matter of finding the "best one" and then always using it. It is the kind of considerations we have just discussed that makes mathematical calculations necessary in computer science and makes computer science an interesting field.

References

Cole, H. S. D. , et al. 1973. *Models of Doom: A Critique of the Limits to Growth.* New York: Universe Books.

Coxeter, H. S. M. 1961. *Introduction to Geometry.* New York: John A. Wiley and Sons, Inc.

Cozzens, M., and R. Porter. 1989. *Recurrence Equations - Counting Backwards.* HiMAP Module 2. Arlington, Massachusetts: COMAP, Inc.

Gleick, J. 1987. *Chaos: Making a New Science.* New York: Viking.

Hofstadter, D. Nov. 1981. Metamagical themas. *Scientific American,* Vol. 245, No. 5, p. 22.

Kolman, B., and R. Busby. 1984. *Discrete Mathematical Structures for Computer Science.* Englewood Cliffs, New Jersey: Prentice-Hall Company.

Kruse, R. L. 1987. *Data Structures and Program Design.* Second Edition. Englewood Cliffs, New Jersey: Prentice-Hall Company.

Malkevitch, J., and W. Meyer. 1974. *Graphs, Models, and Finite Mathematics.* Englewood Cliffs, New Jersey: Prentice-Hall Company.

Mandelbrot, B. B. 1977. *Fractals: Form Chance and Dimension.* San Francisco: W. H. Freeman and Company.

Meadows, D., et al. 1972. *Limits to Growth*. New York: Universe Books.

Merritt, S. M. January, 1985. An Inverted Taxonomy of Sorting Algorithms. *Communications of The Association of Computing Machinery*, p. 96-99.

Newsweek. January 19, 1987. The guru who saw a 2000 dow, p. 40.

Niven, I. 1965. *Mathematics of Choice*. New York: Random House.

Purdom, P. W. and C. A. Brown. 1985. *The Analysis of Algorithms*. New York: Holt, Rinehart and Winston.

Roberts, E. S. 1986. *Thinking Recursively*. New York: John A. Wiley and Sons, Inc.

Rohl, J. S. 1984. *Recursion via Pascal*. New York: Cambridge University Press.

Tenenbaum, A., and M. Augenstein. 1986. *Data Structures Using Pascal*, Second Edition. Englewood Cliffs, New Jersey: Prentice-Hall Company.

Thornburg, D. D. 1983. *Discovering Apple Logo*. Reading, Massachusetts: Addison-Wesley.

Winston, P. H. 1984. *Artificial Intelligence*. Reading, Massachusetts: Addison-Wesley.